LONDON MATHEMATICAL SOCIETY STUDENT TEXTS

Managing editor: Professor C.M. Series, Mathematics Institute
University of Warwick, Coventry CV4 7AL, United Kingdom

3 Local fields, J.W.S. CASSELS
4 An introduction to twistor theory: Second edition, S.A. HUGGETT & K.P. TOD
5 Introduction to general relativity, L.P. HUGHSTON & K.P. TOD
7 The theory of evolution and dynamical systems, J. HOFBAUER & K. SIGMUND
8 Summing and nuclear norms in Banach space theory, G.J.O. JAMESON
9 Automorphisms of surfaces after Nielsen and Thurston, A. CASSON & S. BLEILER
11 Spacetime and singularities, G. NABER
12 Undergraduate algebraic geometry, MILES REID
13 An introduction to Hankel operators, J.R. PARTINGTON
15 Presentations of groups: Second edition, D.L. JOHNSON
17 Aspects of quantum field theory in curved spacetime, S.A. FULLING
18 Braids and coverings: selected topics, VAGN LUNDSGAARD HANSEN
19 Steps in commutative algebra, R.Y. SHARP
20 Communication theory, C.M. GOLDIE & R.G.E. PINCH
21 Representations of finite groups of Lie type, FRANÇOIS DIGNE & JEAN MICHEL
22 Designs, graphs, codes, and their links, P.J. CAMERON & J.H. VAN LINT
23 Complex algebraic curves, FRANCES KIRWAN
24 Lectures on elliptic curves, J.W.S. CASSELS
25 Hyperbolic geometry, BIRGER IVERSEN
26 An introduction to the theory of L-functions and Eisenstein series, H. HIDA
27 Hilbert Space: compact operators and the trace theorem, J.R. RETHERFORD
28 Potential theory in the complex plane, T. RANSFORD
29 Undergraduate commutative algebra, M. REID
31 The Laplacian on a Riemannian manifold, S. ROSENBERG
32 Lectures on Lie groups and Lie algebras, R. CARTER, G. SEGAL & I. MACDONALD
33 A primer of algebraic *D*-modules, S.C. COUTINHO
34 Complex algebraic surfaces, A. BEAUVILLE
35 Young tableaux, W. FULTON
37 A mathematical introduction to wavelets, P. WOJTASZCZYK
38 Harmonic maps, loop groups and integrable systems, M. GUEST
39 Set theory for the working mathematician, K. CIESIELSKI
40 Ergodic theory and dynamical systems, M. POLLICOTT & M. YURI
41 The algorithmic resolution of diophantine equations, N.P. SMART
42 Equilibrium states in ergodic theory, G. KELLER

London Mathematical Society Student Texts 41

# The Algorithmic Resolution of Diophantine Equations

Nigel P. Smart
Hewlett-Packard Laboratories, Bristol

CAMBRIDGE
UNIVERSITY PRESS

CAMBRIDGE
UNIVERSITY PRESS

University Printing House, Cambridge CB2 8BS, United Kingdom

Cambridge University Press is part of the University of Cambridge.

It furthers the University's mission by disseminating knowledge in the pursuit of education, learning and research at the highest international levels of excellence.

www.cambridge.org
Information on this title: www.cambridge.org/9780521646338



First published 1998

*A catalogue record for this publication is available from the British Library*

*Library of Congress Cataloguing in Publication data*

Smart, N. P. (Nigel Paul), 1967–
The algorithmic resolution of diophantine equations / N.P. smart
p. cm.
Includes bibliographical references and index.
ISBN 052164156X. – ISBN 0521646332 (pbk.)
1. Diophantine equations. I. Title.
QA242.S69 1998
512'. 72–dc21 98–24736 CIP

ISBN 978-0-521-64156-2 Hardback
ISBN 978-0-521-64633-8 Paperback

To Maggie, Eleanor and Oliver.

# Contents

# Preface

Many books have been devoted to the theoretical study of diophantine equations, an observation which should come as no surprise given that the study of such equations dates back over two thousand years. In theoretical work one is interested in determining the structure of the solution set to some equation. Is the set finite or infinite? Can one give an effective procedure to determine all the solutions? Do the solutions form a group of some sort? How are the rational solutions distributed amongst the real solutions? The list of questions that one can ask is endless.

In this book we shall concentrate on algorithms and methods for writing down all the solutions to an equation (if there are finitely many) or for determining explicitly the structure of all of the solutions (if there are infinitely many). Despite the long and noble career of diophantine equations, there appear to be only two books solely devoted to the study of explicit methods for their solution, namely Mordell's *Diophantine Equations* [**138**] and de Weger's *Algorithms For Diophantine Equations* [**208**].

Mordell's book gives a variety of techniques for solving various diophantine equations. However, sometimes he deals just with special cases and sometimes with general cases. Mordell does not concentrate on algorithmic questions and hence some of his methods appear at first sight to be recipes which only apply to certain special cases. This is not surprising as it was originally published in 1969, before the advent of the modern desktop computer.

The second book is de Weger's thesis, in which the systematic use of the LLL–algorithm was proposed for solving diophantine equations. This has revolutionized the subject and led to a great explosion in the number of papers devoted to algorithms for diophantine equations. De Weger's book was published in 1989 at the beginning of this revolution and it therefore only barely touches, for instance, on the algorithm for Thue–Mahler equations developed by Tzanakis and de Weger.

There have been many books which have studied diophantine equations from a theoretical standpoint, most notably the book by Shorey and Tijdeman [**167**], which gives an excellent account of the applications of Baker's theory of linear forms in logarithms.

The advent of modern computer technology has led to a number of books on algorithms for number theory. We shall require the use of various algorithms to solve problems in algebraic number theory. In particular we shall require the solution of various problems in algebraic number fields, such as unit and class group computation. Many of the number field algorithms we require can be found in the books by Cohen [**32**], Pohst [**154**] and Pohst and Zassenhaus [**155**]. For up to date information one should perhaps consult the various conference proceedings, such as [**1**], [**33**], [**137**] and [**152**].

Therefore the time seems ripe for a new book on the computational side of this area. We shall aim to provide a coherent account of some of the many methods that can be used to find all the solutions to certain diophantine equations. However, we shall mainly be interested in methods which apply to a wide class of equations rather than just a few special examples. In some sense this is still a recipe book, but we hope a recipe book which gives the chef a range of skills for coping with a number of dishes.

We shall assume that you are familiar with standard undergraduate algebraic number theory up to, say, Dirichlet's units theorem and the finiteness of the class group. The book by Rose [**160**] covers most of what we will require bar the two aforementioned results on the units and class number. For these last two results you should perhaps consult another textbook such as that by Stewart and Tall [**187**].

We shall also assume that you have begun to study the more advanced theory that one meets as a graduate student, such as local fields. We shall, however, give a brief overview of the theory of local fields at the start. We shall furthermore take it that the reason you are reading this book is that you are interested in computations in number theory. This is not, therefore, a theoretical book but a practical one.

## Outline

The book is divided into three parts. Part 1 will involve the study of the basic techniques which are used over and over again in solving diophantine equations. These are chiefly: the theory of $p$-adic numbers; the use of curves of genus zero; and the application of the algorithm of Lenstra, Lenstra and Lovász.

We shall, in Chapter II, start by giving a brief overview of the theory of $p$-adic numbers and local fields. Those of you who have not met local fields before should consult one of the excellent textbooks in this area. This subject is covered very well in other books such as the one by Cassels, [**24**], so we shall just review the main results we shall need. An explanation of $p$-adic analogues of results from numerical analysis will start our investigation.

In Chapter III we shall focus on how we can use the theory of $p$-adic numbers to solve diophantine equations, with the use of Hensel's lemma and Skolem's method. Finally we shall end our discussion of local fields with a brief discussion on how one can put all the local information together in various ways using Hasse's principle and sieving. The discussion of these applications of local methods will be brief, as they are covered elsewhere, for instance in the book by Mordell [**138**] mentioned previously.

We shall then turn, in Chapter IV, to the discussion of the solution of ternary quadratic forms. These can also be characterized as curves in $\mathbb{P}^2$ of genus zero. Such equations are not only important in their own right but also occur in algorithms for solving more complicated equations, the idea being that ternary quadratic forms are 'easy' and if we can reduce our study of a hard equation to a set of ternary quadratic forms then we would have made life easier. The reason ternary quadratic forms are considered easy is that they all satisfy the Hasse principle.

In Chapter V we study the LLL–algorithm of Lenstra, Lenstra and Lovász. We shall develop the algorithm from scratch and go on to show how one can use it to give a lower bound on the size of the smallest non-zero vector in a lattice and the smallest distance between a given non-lattice vector and a vector in the lattice.

Chapter VI will be concerned with the application of the LLL–algorithm to various problems. For instance we shall see how one can use LLL to solve certain types of knapsack problem. We shall end by showing how one can use LLL to study problems of linear forms in real, complex and $p$-adic numbers. This technique is the one we shall use to reduce the stratospherically large bounds which arise from the theory of linear forms in logarithms.

Part 2 will be devoted to problems to which one can apply the theory of linear forms in logarithms of algebraic numbers and its generalizations to $p$-adic logarithms.

Thue and Thue–Mahler equations are dealt with in Chapters VII and VIII. Thue equations are equations of the form

$$F(X,Y)=m$$

whilst Thue–Mahler equations are of the form

$$F(X,Y)=mp_1^{z_1}\cdots p_t^{z_t},$$

where, in both cases, $F(X,Y)$ is a binary form of degree greater than three and $m$ is some fixed integer. But in the case of Thue–Mahler equations we have the added complication of some given prime numbers $p_i$ with some unknown exponents $z_i$. Thue equations have a finite number of integer solutions, whilst Thue–Mahler equations have infinitely many solutions which can be divided into finitely many families. Thue and Thue–Mahler equations form the easiest examples of classes of equations which can be dealt with by Baker's methods followed by a reduction process based on the LLL–algorithm. Chapter VII will conclude with an example of how to use Thue equations to solve another diophantine problem: that of finding all the integer points on an elliptic curve. This will be the first of three such methods we give to solve this problem.

The trick in solving Thue and Thue–Mahler equations is to reduce the problem to the study of $S$-unit equations, that is, equations of the form

$$\alpha_1\tau_1 + \alpha_2\tau_2 + 1 = 0,$$

where $\alpha_i$ are two fixed algebraic numbers and $\tau_1$ and $\tau_2$ are allowed to range over two finitely generated multiplicative subgroups of the algebraic numbers. Such an equation has only finitely many solutions, which we shall give an effective proof of in Chapter IX. Indeed we shall give an algorithm which can often be used in practice to solve such an equation. We end this chapter by showing how one could use an algorithm to solve $S$-unit equations to give another method for finding all the integral points on an elliptic curve.

In that late 1970s and early 1980s Győry showed how a very large set of diophantine equations could be reduced to the study of $S$-unit equations. This set of triangularly connected decomposable form equations (TCDF equations for short) is studied in Chapter X. These equations are a natural generalization of the Thue and Thue–Mahler equations considered earlier.

In Chapter XI we shall pay particular attention to a special type of TCDF equation, the set of discriminant form equations. We shall end this chapter by showing how discriminant form equations related to quartic number fields can be solved by using a combination of Thue equations and ternary quadratic forms, which bypasses the need to consider them as TCDF equations.

In Part 3 we shall consider methods for finding integral and rational solutions to curves such as elliptic, hyperelliptic and superelliptic equations.

In Chapters XII and XIII we shall concentrate on elliptic curves. It has been known for over 100 years that the set of rational points on an elliptic curve forms a group. Chapter XII will be devoted to giving an (almost) algorithmic proof of the result of Mordell that such a group is finitely generated. That there is no such algorithmic answer in general is due to the failure of the Hasse principle for curves of genus one. In Chapter XIII we shall use the method for determining generators of the group of rational points to give a third method for finding all the integral points on such a curve.

In Chapter XIV we shall look at recent work on generalizations of the methods for elliptic curves to curves of higher genus. In particular we shall concentrate on hyperelliptic curves. Owing to Faltings proof of the Mordell conjecture we now know that there are only finitely many rational points on a curve of genus greater than one. However, at present there are only ad hoc techniques to find all the rational points in any given example. We shall present a quick overview of some of the work done in this area and its link with the Jacobian variety of a curve of genus greater than one.

In this final chapter we shall also cover a few odds and ends which we have not covered in other chapters. In particular no book on diophantine equations would be complete without a passing mention of Wiles' proof of Fermat's last theorem. In this last section we shall describe the link between Fermat's last theorem and elliptic curves, although we shall not go into any details, as that would involve going into the theory of modular functions and Galois representations. In addition in this last chapter we shall look at the $ABC$ conjecture, which is in some sense a generalization of the two–term $S$-unit equations which are met elsewhere in the book.

Clearly we have not even attempted to cover all the different types of equation that can be studied. Nor have we covered much of the extensive theoretical work on diophantine equations. The subjects chosen are a personal choice, as is fitting for a recipe book. There are some topics which we have left out owing to lack of space.

One is the application of Padé approximations and hypergeometric functions. In this work instead of trying to approximate linear forms in logarithms one looks at, for example, approximating numbers of the form

$$\sqrt[n]{1+a}.$$

Readers interested in following up such work should consult [**159**], [**29**], [**120**], [**206**], [**8**] and [**9**].

Another is the algorithmic study of diophantine properties of linear recurrence sequences. If this area is what interests you then why not start by looking at [**145**] and [**214**]. A good introduction to this area can be found in the relevant chapters of [**167**].

## Computer packages

There are currently many computer packages for performing number theoretic calculations. We could be content with just using one of the main computer algebra packages such as **Maple** or **Mathematica**. However, we shall need to be able to compute units and class groups of number fields, etc. Hence access to a package like **PARI** [7], **KANT** [42], **SIMATH** [177], **LiDIA** [122] or **MAGMA** [17], would seem desirable.

Many of the examples in this book were carried out with the aid of a computer, so you should not expect to be able to follow an example through by hand (except in some easy cases). However, a computer can solve most of the examples in this book in a matter of seconds.

## Notation

As usual we shall denote the complex, real and rational numbers by $\mathbb{C}, \mathbb{R}$ and $\mathbb{Q}$. The ring of integers we shall denote by $\mathbb{Z}$ while the set of non-negative integers will be denoted by $\mathbb{N}$. Multiplication of numbers will denoted by $6 = 2 \cdot 3$, while a decimal point will be given by $2.3 = \frac{23}{10}$.

Some of the notation used could be considered non-standard in that not all authors use the same notation. To make this clear we spell out the possible non-standard notation now.

The notation $\mathbb{Z}_p$ will be reserved for the $p$-adic integers, the $p$-adic numbers being denoted by $\mathbb{Q}_p$. The set of integers modulo $m$ will then be denoted by $\mathbb{Z}/m\mathbb{Z}$. The finite field of $q$ elements will be denoted by $\mathbb{F}_q$

For a real number $x$, the symbol $\lfloor x \rfloor$ will denote the floor function, i.e. it returns the largest integer less than $x$. The symbol $\lceil x \rceil$ will denote the ceiling function, i.e. the smallest integer greater than $x$. The nearest integer function will be denoted $[x]$, with any fixed convention for numbers of the form $(2m+1)/2$. The symbol $\{x\}$ will be used to denote $|x - [x]|$.

For a complex number $z$ the real and imaginary parts will be denoted by $\Re(z)$ and $\Im(z)$.

The symbol ${}^nC_r$ will denote the binomial coefficient

$$\frac{n!}{r!(n-r)!}.$$

The greatest common divisor of two integers $a$ and $b$ will be denoted by $(a, b)$.

If $K$ is a number field then we let $\mathcal{O}_K$ denote its maximal order. The unit and class groups of $K$ will be denoted by $\mathcal{O}_K^*$ and $CL_K$ respectively. If $\alpha_1, \dots, \alpha_t$ are elements of $\mathcal{O}_K$, for some number field, $K$ then we let $(\alpha_1, \dots, \alpha_t)$ denote the ideal generated over $\mathcal{O}_K$ by $\alpha_1, \dots, \alpha_t$.

All other notation will be defined as and when required.

## Thanks

The author would like to thank J. Cremona, S. Crouch, E. Mansfield, A. Pethő, R. Shipsey, S. Siksek, N. Stephens and B. M. M. de Weger, who read various parts of the manuscript at various stages. Any mistakes are still, however, my own fault. None of the mathematics in this book is new and an attempt has been made to provide references to major results. If you feel that due credit has not been given for certain results, then accept the authors apologies in advance.

Finally, thanks are due to J. Cremona for a TeX macro for typesetting the algorithms.

CHAPTER I

# Introduction

This book shall concern itself with the study of modern methods for solving diophantine equations. The study of diophantine equations goes back to the ancient Greeks. The most famous example from that time, $X^2 + Y^2 = Z^2$, is still being taught in schools today. Many of the ideas in this book can be traced back to earlier times, so I shall start by giving a brief outline of the history of the subject. This will be to set the scene and raise the problems that will hopefully be answered in the following chapters.

By a diophantine equation we mean, intuitively, an equation where we are interested only in integer or rational solutions. For example, Fermat's famous 'last theorem', now Wiles' theorem, says that the only integer solutions to the equation

$$X^n + Y^n = Z^n \tag{I.1}$$

with $n \geq 3$ are given by $XYZ = 0$. Another important class of examples is elliptic curves, which are curves of the form

$$y^2 + a_1xy + a_3y = x^3 + a_2x^2 + a_4x + a_6.$$

When studying equation (I.1) it clearly makes no difference if we study the rational solutions $(X, Y, Z)$ or the integral solutions. However, when looking at elliptic curves it makes a great deal of difference whether we want to determine the rational or integral solutions. An elliptic curve can (and often does) possess an infinite number of rational solutions, but it will only ever possess a finite number of integral solutions, as we shall see in a later chapter.

Factoring an integer can be considered as solving a diophantine equation. Given an integer $N$ the problem of factoring can be presented as finding the integral solutions to the equation

$$N = pq$$

where $p, q \in \mathbb{N}$.

Diophantine equations have over the centuries provided a fertile ground for mathematical investigation. This is at first glance surprising, as finding solutions to an equation in the real numbers appears easy. We can, for instance, just draw a graph, and the integers are considered a much simpler mathematical object than the reals.

## I.1. A brief history

The study of diophantine equations dates back to at least 1600BC. The earliest work of importance seems to have been on the problem of determining 'Pythagorean' triples, that is, non-trivial solutions $(X, Y, Z)$ to the equation

$$X^2 + Y^2 = Z^2.$$

Any school child knows about the triple $(3, 4, 5)$, while any undergraduate would understand the proof that all triples are given by (up to an interchange of $x$ and $y$)

$$\begin{aligned} x &= \pm d(a^2 - b^2), \\ y &= \pm 2abd, \\ z &= \pm d(a^2 + b^2), \end{aligned}$$

where $a, b, d \in \mathbb{Z}$ with $\gcd(a, b) = 1$.

The name diophantine equations is in honour of the mathematician Diophantos, who lived in Alexandria around 300AD. Diophantos' work *Arithmetica* was one of the ancient texts that went 'missing' in Europe in the Dark and Middle Ages. The *Arithmetica* originally consisted of 13 books, of which only 6 have survived into the modern era. Two translations of the remaining books were made in the sixteenth and seventeenth centuries. It was in the margin of Fermat's copy of Bachet's translation of *Arithmetica* that Fermat made his famous marginal note that equation (I.1) has no non-trivial solutions.

Pierre de Fermat (1601–65) gave a large number of legacies to mathematics, and in particular number theory, the most famous of these being the above–mentioned last theorem. More important was his introduction of the so called 'method of descent'. In this method one supposes one has a solution which is as 'small' as possible, and then one produces by some means an even 'smaller' solution. This contradiction tells us that our original solution could not have existed in the first place. Fermat applied his method of descent to show

THEOREM I.1 (Fermat). *The equation*

$$z^2 = x^4 + y^4$$

*has no non-trivial integer solutions.*

PROOF. Suppose that there is a non–trivial solution. We can clearly assume that it satisfies $(x, y) = 1$ and without loss of generality we can assume that $x$ is odd, $y$ is even and both are positive. By applying the formulae for pythagorean triples given above, we can then write

$$x^2 = a^2 - b^2\ ,\ y^2 = 2ab\ ,\ z = a^2 + b^2,$$

for two coprime integers $a$ and $b$. We then apply the formulae again to the equation $x^2 + b^2 = a^2$ to obtain

$$a = p^2 + q^2 \ , \ x = p^2 - q^2 \ , \ b = 2pq,$$

where $p$ and $q$ are two coprime integers. Since $y$ is even we obtain

$$\left(\frac{y}{2}\right)^2 = \frac{ab}{2} = pq(p^2 + q^2),$$

which leads us deduce, as $p, q$ and $p^2+q^2$ are coprime, that there exist positive integers $X, Y$ and $Z$ such that

$$p = X^2 \ , \ q = Y^2 \ , \ p^2 + q^2 = Z^2,$$

and so $(X, Y, Z)$ is another solution to

$$Z^2 = X^4 + Y^4.$$

To sum up, we have from one solution to our equation deduced another solution to our equation. The trick of the proof is to show that this new solution is 'smaller' than the original one. If we can do this then this method of descending to a 'smaller' solution cannot be carried on indefinitely, and so the original solution could not have existed in the first place.

Note that

$$y = 2XY\sqrt{X^4 + Y^4},$$

so if either $X$ or $Y$ is zero then $y$ is also zero, which would mean that $(x, y, z)$ was a trivial solution. Hence neither $X$ nor $Y$ can be zero. It is then clear that $X < y$ and $Y < y$. So the new solution must be 'smaller' than the old solution. □

As a corollary we easily deduce that Fermat's last theorem holds for the exponent $n = 4$. The method of descent has since been adapted and now the name 'descent' is often given to any process whereby the existence or non-existence of solutions to some equation is proved by means of considering other, in most cases smaller, solutions to either the same equation or a related set of equations. Using the method of descent it is believed that Fermat managed to show that if $p$ is a prime congruent to 1 modulo 4 then the equation

$$x^2 + y^2 = p$$

always has an integer solution. However, no proof of this result by Fermat survives; the earliest known proof dates back as far as Euler. This descent method uses a known solution to one equation to deduce a solution to a similar equation with smaller coefficients:

THEOREM I.2. *Let $p$ denote a prime congruent to one modulo four. Then there exists a solution in integers to the equation*

$$p = x^2 + y^2.$$

PROOF. Clearly there is a solution $(x_1, y_1)$ to the equation

$$mp = x_1^2 + y_1^2$$

for some positive integer value of $m$. For example we can take $x_1$ to denote a square root of $-1$ modulo $p$ and $y_1$ to be 1. With this choice we can assume that $m < p$. Now choose two integers $(u, v)$ such that $u \equiv x_1 \pmod{m}$, $v \equiv y_1 \pmod{m}$ and $u, v \in [-m/2, m/2]$. We then have that

$$u^2 + v^2 \equiv x_1^2 + y_1^2 \pmod{m},$$

so that

$$u^2 + v^2 = rm$$

for some positive integer value of $r < m$. Now set

$$x_1' = x_1 u + y_1 v \ , \ y_1' = x_1 v - y_1 u.$$

Then we notice that both $x_1'$ and $y_1'$ are multiples of $m$. So we set $x_1' = mx_2$ and $y_1' = my_2$. But then

$$\begin{aligned} m^2(x_2^2 + y_2^2) &= x_1'^2 + y_1'^2 = (x_1 u + y_1 v)^2 + (x_1 v - y_1 u)^2 \\ &= (x_1^2 + y_1^2)(u^2 + v^2) \\ &= rpm^2. \end{aligned}$$

Hence $(x_2, y_2)$ is a solution to the equation

$$x_2^2 + y_2^2 = rp,$$

where $r < m$. We can continue carrying out this process, but not indefinitely, as the values of $r$ are positive and get successively smaller. Hence at some point we will reach $r = 1$ and we will have a solution to our equation. □

Euler and Lagrange also gave a proof of a result which had been asserted by Fermat, namely that every integer could be written as the sum of four rational squares. Fermat had claimed he had a proof of this result which also used his method of descent.

Our story now jumps forward a century to the time of Gauss, who was born in Braunschweig in 1777. Gauss has obtained the reputation of being one of the most original mathematicians in history. Up to his death, in 1855, he worked in various areas of mathematics and physics such as algebra, magnetism and probability. However, it is his work in number theory which interests us. Gauss studied the integer solutions to quadratic equations such as

$$ax^2 + bxy + cy^2 + dx + ey + f = 0,$$

where $a, b, c, d, e, f$ are given integers. This includes the case of Pell's equation

$$y^2 - Dx^2 = 1,$$

which provides a useful motivating example in undergraduate courses for the topics of quadratic fields and continued fractions. Some examples of Pell's equation had even been solved in Diophantos's *Arithmetica*, and some had been studied by the Hindu mathematicians Brahmagupta and Bhaskara in the seventh and twelfth centuries respectively. It was Euler who first spotted the link between solutions of Pell's equation and the continued fraction expansion of the quadratic irrational $\sqrt{D}$, a link which is sometimes used today to compute fundamental units of real quadratic number fields. Gauss developed the arithmetic of quadratic number fields, in particular their class groups. However, he expressed everything in terms of the theory of binary quadratic forms. The development of the notion of ideals was to come after Gauss.

The challenge of Fermat's last theorem led, amongst other problems, to the development in the nineteenth century of the subject known as algebraic number theory. This subject provided the testing ground for much of modern algebra. The development of the notions of rings, ideals, modules, unique factorization domains and other basic notions can be traced back to the investigation of algebraic number fields. The list of mathematicians involved in this theory, Dedekind, Dirichlet, Galois, Kummer, Minkowski, etc., can be recounted by any undergraduate number theorist.

Another interest of Fermat was the study of elliptic curves. That the rational points form a group has been known for over a century, as the group law can be deduced from the classical chord–tangent process, a process linked to the addition formulae for elliptic integrals. Weierstrass (1815–97) had studied such elliptic functions and had expanded on the work of the Norwegian mathematician Abel (1802–29). Abel had discovered the class of transcendental functions which we now call Abelian functions, of which elliptic functions are an example. Earlier Gauss had seen the link between elliptic integrals and his arithmetic–geometric mean, but his work in this area was not published until after his death (and after the publication of Abel's work).

A similar process to the chord–tangent process had been used by Fermat to deduce more solutions to an elliptic curve given a known solution, this being known as Fermat's method of ascent. The name ascent comes from the fact that the process, in general, produces larger and larger rational solutions from known smaller ones.

Jules Henri Poincaré (1854–1912), although more famous for his works in topology and mathematical physics, in the late 1800s conjectured that the set of rational points on any given elliptic curve, including an additional point 'at infinity', formed a finitely generated abelian group. In 1922 Mordell proved Poincaré's conjecture using a technique based on Fermat's method of descent.

The first half of the twentieth century also saw the introduction of 'local methods' into number theory by Hasse, Hensel, Skolem and others. These were used, especially by Skolem, to find all the solutions to a large number of individual equations. Indeed Skolem's method was still the main one in use up to the late 1970s, and it is still of relevance today. Later, in Chapter III, we shall give some examples of the use of Skolem's method to solve equations.

In 1913, Ramanujan asked what are the positive integral solutions to the diophantine equation

$$x^2 + 7 = 2^n.$$

In 1948 Nagell showed that there are only five positive integral solutions given by

$$(n, x) = (3,1), (4,3), (5,5), (7,11), (15,181).$$

This equation is now called the Ramanujan–Nagell equation. A proof that these are the only five positive solutions of the Ramanujan–Nagell equation can be found in many standard textbooks, such as [**187**]. Various authors have considered generalizations of the Ramanujan–Nagell equation, the most general form being that considered by Pethő and de Weger,

$$x^2 + k = p_1^{z_1} \cdots p_t^{z_t}.$$

In [**151**] a general method is given to deal with this generalized Ramanujan–Nagell equation. This shows a marked change of emphasis, from the study of individual equations to the study of generalized classes of equations.

Prior to the Second World War, if the solutions to an equation were able to be determined explicitly then usually some special trick was used which only applied to that equation. If you wanted to find the solutions to a similar equation, you had to find a similar trick which worked in this new case. Very few methods, Skolem's being the notable exception, could be applied to a large number of equations without alteration.

Given a diophantine equation, or even a class of equations such as 'all elliptic curves', there are some natural questions that come to mind:

1. Is the number of rational (or integral) solutions finite or infinite?
2. If the number of rational (or integral) solutions is finite, can we give a procedure which in a finite amount of time will determine all the solutions?
3. If the number of rational (or integral) solutions is infinite, is it possible to express all solutions in terms of some 'basic' ones? For instance, given Mordell's result that the rational points on an elliptic curve form a finitely generated group, can we construct explicit generators for this group in any given example?

Answering either of the final two questions clearly is a harder task than the first one.

If we can prove that an equation has a finite number of solutions in a way which gives an algorithm to determine all the solutions, then one is said to have an **effective** proof. If on the other hand one can only answer the question of the finiteness of the number of solutions without giving an algorithm, then one is said to have an **ineffective** proof. Hilbert, in the tenth of his famous problems at the turn of the century, asked whether there existed an algorithm which could determine whether any diophantine equation had finitely or infinitely many solutions.

In 1970 Matijasevič proved the non-existence of such an algorithm. See [**45**] for a general discussion on Hilbert's tenth problem and Matijasevič's solution. However, this still left the question open as to whether large classes of equations could be tackled using algorithms.

Three years before Matijasevič's proof Baker [**4**] had given an effective proof of a result of Thue. Thue [**196**] had shown in an ineffective way that there are only finitely many integer solutions to equations of the form

$$F(X, Y) = m,$$

where $F(X, Y)$ is a binary form of degree greater than 2 and $m$ is some fixed integer, for example, equations such as

$$X^3 + 2XY^2 + Y^3 = 2$$

or

$$X^{16} - 33Y^{16} = 42775.$$

Such equations are now called Thue equations. Baker showed how one could bound the size of the solutions in terms of the coefficients of $F$ and the integer $m$. Hence there is a finite search region within which all the solutions must lie. However, the bounds given by Baker are huge and certainly not meant for a practical solution of Thue equations.

As Baker's method results in a finite search region, it was not long before people started thinking of ways of reducing the size of this region. This was began in the early 1970s with work of Baker, Davenport and Ellison, see [**5**] and [**49**] . However, in the mid–1980s after the development of the LLL–algorithm by Lenstra, Lenstra and Lovász [**117**] a new technique was available. This technique, which was pioneered in de Weger's thesis [**208**] allowed the computational resolution of equations which only a few years previously had been the preserve of theoreticians.

## I.2. Algorithms

In this book we shall mainly be concerned with effective proofs. But we shall also be interested in techniques which not only tell us how we could compute all the solutions to an equation in principle but how we could do so in **practice**. Consider the following example. Suppose one could show that all integral solutions to the Thue equation

$$X^7 + 2Y^7 = 1$$

satisfied

$$|X|, |Y| \leq 10^{40}.$$

We would then know that there were finitely many solutions. Not only that but we could give a computer the task of computing all the solutions. If the computer could check whether a pair $(X, Y)$ was a solution in, say, one billionth of a second, then it would take the computer around $10^{50}$ years to go through all the possible solutions to the equation. Unfortunately this is much longer than the estimated age of the Universe.

Therefore we cannot be content only with effective methods. What will interest us is methods which can be applied in practice to actually determine all the solutions to a problem. Even here one has problems, as an algorithm which will determine all solutions in one example may not work in another. It could fail either because the algorithm does not apply or because the algorithm would take too long to be of any practical benefit.

We shall be interested in practical algorithms which apply to wide classes of equations and we will be satisfied if the algorithm works for a wide number of examples in the class. We shall not discuss the problems of growth of expected running time and other complexity theoretic issues. Indeed the complexity theoretic study of algorithms for diophantine equations is not anywhere near as well developed as it is for other areas in computational number theory, but we shall need an intuitive concept.

Upper bounds on run times of algorithms depend, of course, on the definition one is using for time. The standard definition is to measure time in terms of bit operations; this is often referred to as 'bit complexity'. We shall not really worry about the exact definitions from complexity theory, but we will have a need to measure and compare one algorithm's expected run time against another. For this comparison we will be content with just an intuitive understanding; for a more complete discussion of such matters you should consult [**3**].

We shall sometimes refer to an algorithm as running in polynomial or exponential time in some parameter. We shall now review this concept briefly, as it may be new to some readers. An algorithm is said to run in polynomial time, with respect to some parameter $B$, if its run time is bounded by a polynomial function of $B$; in other words its run time is $O(B^n)$ for some fixed number $n$. An algorithm is said to run in exponential time if it runs in time $O(n^B)$. An algorithm is said to run in subexponential time if its run time is certainly faster than exponential in behavior but not necessarily as good as polynomial in growth.

Normally the run time is measured in terms of the length of the input data. So an algorithm which has as input a single number $N$ is said to run in polynomial time if we can bound the run time by $O((\log N)^n)$, for some fixed number $n$, while an algorithm runs in exponential time if we can bound the run time by $O(N^n)$, for some fixed $n$. This is because the length of the input to the algorithm is of size $O(\log N)$.

It is convenient to introduce the estimate

$$L_N(\alpha, \beta) = O\left( (e^{(\log N)^\alpha (\log\log N)^{1-\alpha}})^{\beta + o(1)} \right).$$

This interpolates between polynomial time, $\alpha = 0$, and exponential time, $\alpha = 1$. This estimate occurs quite a lot in number theoretic algorithms. For example the best–known factoring algorithm has complexity $L_N(1/3, c)$ for some constant $c$, while the best–known algorithms for computing the class group and unit group of a number field have complexity $L_D(1/2, c)$, where $D$ is the discriminant of the field and one assumes the generalized Riemann hypothesis (GRH).

In this book three main techniques will be discussed for solving diophantine equations, being:

- The application of 'local' considerations. This includes Skolem's method and sieving.
- Reducing the problem of finding the solutions to one equation to the problem of finding the solutions to another set of hopefully easier equations. We shall meet this, for instance, when we discuss the method of descent for elliptic curves.
- The application of Baker's theory of linear forms in logarithms and the use of a method to reduce the huge bounds resulting from this theory.

Baker's technique itself divides into four main steps:

1. Reduce the solution to a problem for which we know Baker's techniques will apply.
2. Produce effective bounds on solutions using deep theoretical results of Baker, Yu, Waldschmidt, Wüstholz and others.
3. Reduce these large bounds to something more manageable, using a computational technique first developed by de Weger.
4. Find clever search techniques to find all solutions under the reduced bounds.

In recent years other techniques from areas such as arithmetic geometry have been developed. These allow one, for instance, to find all rational points on some curves of genus greater than one. That there are finitely many such rational points follows from the deep, but ineffective, work of Faltings. In later sections we shall consider these new methods and ideas and current open problems.

Many algorithms to find all the solutions to a particular equation require fast and efficient techniques for finding one (or a few) solutions to another set of equations. Sometimes we need to find all solutions below a certain upper bound. To solve this problem we will use a 'sieving' technique which can be found in various guises throughout this book. Often it is the sieve, which locates all small solutions or exhibits a single solution, which is the slowest part of the entire solution process.

## I.3. What is a diophantine equation?

Often the above definition of a diophantine equation is too restrictive. We would like to consider generalizations of the intuitive notion of diophantine equation considered above. We shall now define the exact notion which shall be used in this book. What was important in our intuitive definition was that a diophantine equation was not only the equation but also the set for which were trying to find solutions in. In the examples above this set was always $\mathbb{Z}$ or $\mathbb{Q}$. It makes sense to admit other sets as candidates for containing solutions.

Let $K$ denote an algebraic number field and let $S_i$ denote some well–defined subsets of $K$, for $i = 1, \ldots, n$. If $F(X_1, \ldots, X_n)$ denotes some function which maps $S_1 \times \cdots \times S_n$ to $K$, then we define a diophantine equation to be the equation

$$F(X_1, \ldots, X_n) = 0,$$

where we are interested in determining the structure of all the solutions in $S_1 \times \cdots \times S_n$.

There is a common equivalence we use for inhomogenous equations in two variables or homogenous equations in three variables. Such equations are usually thought of as curves, so we often speak of 'points on the curve' as meaning 'solutions to the equation'. Clearly our solutions (and hence points) can come from various sets, the term 'integral point' being reserved for points which have coordinates in the ring $\mathbb{Z}$.

For instance, we can use this to define what it means for an elliptic curve defined over a number field to have integral points. Let

$$Y^2 = X^3 + AX + B$$

denote an elliptic curve defined over $K$, by which we mean $A, B \in K$. Then determining all the solutions in $\mathcal{O}_K \times \mathcal{O}_K$ to the equation

$$F(X,Y) = Y^2 - X^3 - AX - B = 0$$

is what we mean by finding all the integral points on the elliptic curve, in this example $S_1 = S_2 = \mathcal{O}_K$. If we replace $S_1$ and $S_2$ by $K$, we are then faced with the task of determining the structure of the $K$-rational points on the elliptic curve.

In this vein one of the most important types of equation we shall meet is the two–term $S$-unit equations. In these equations we let $S_1$ and $S_2$ denote two finitely generated subgroups of $K^*$ and let $\alpha_1, \alpha_2$ denote two fixed elements of $K^*$. By solving a two–term $S$-unit equation we mean determining all the solutions to the following equation, with $X_1 \in S_1$ and $X_2 \in S_2$:

$$F(X_1, X_2) = \alpha_1 X_1 + \alpha_2 X_2 + 1 = 0.$$

## I.4. An elliptic curve

We end this introduction with an example of a diophantine equation which can be solved using standard techniques from undergraduate number theory. We shall then discuss just how lucky we are in this situation and the type of problems which can occur when one carries out the following ideas for other examples.

We look at the problem of finding all the integer solutions to a specific example of an elliptic curve. The example we have chosen is a standard one which occurs in many textbooks. We shall use the fact that the ring of integers of $\mathbb{Q}(\sqrt{-1})$, which is $\mathbb{Z}[\sqrt{-1}]$, is a principal ideal domain and hence is a unique factorization domain.

THEOREM I.3. *The only integer solutions to the equation*

$$Y^2 = X^3 - 4$$

*are given by* $(X, Y) = (5, \pm 11)$ *and* $(2, \pm 2)$.

We first prove a couple of preliminary lemmas, which helps break the argument up:

LEMMA I.4. *In* $\mathbb{Z}[i]$ *the two numbers* $2 + iY$ *and* $2 - iY$ *are coprime, for any odd number* $Y$.

PROOF. Suppose the two numbers are divisible by $a + bi$, where $a, b \in \mathbb{Z}$, then so is their sum and difference. Hence $a + bi$ divides 4 and $2iY$. Taking norms we see that $a^2 + b^2$ divides 16 and $4Y^2$. Hence $a^2 + b^2$ divides 4 as $Y$ is odd. But we also have that $a^2 + b^2$ divides the norm of $2 + iY$, which is $4 + Y^2$. Hence $a^2 + b^2$ divides $Y^2$. The only possibility is that $a^2 + b^2 = 1$, which means $a + bi$ is a unit, and so $2 + iY$ and $2 - iY$ are coprime. □

LEMMA I.5. *Assume* $1 + i$ *divides one of* $1 + iZ$ *or* $1 - iZ$. *Then the highest common factor of* $1 + iZ$ *and* $1 - iZ$ *in* $\mathbb{Z}[i]$ *is given by* $1 + i$, *for any odd number* $Z$.

PROOF. Suppose $a + bi$ divides both $1 + iZ$ and $1 - iZ$. Then $a + bi$ divides their sum, namely 2. Hence $a^2 + b^2$ divides both 4 and $1 + Z^2$, but as $Z$ is odd we have $a^2 + b^2 = 1$ or 2. This means that $(a, b) = (\pm 1, 0), (0, \pm 1), (\pm 1, \pm 1)$ or $(\pm 1, \mp 1)$. Hence $a + bi$ is either a unit or an associate of $1 + i$. □

PROOF. [of Theorem I.3] We divide the proof into two cases:

$Y$ **odd.** In $\mathbb{Z}[i]$ our equation factorizes as

$$(2 + iY)(2 - iY) = X^3.$$

By Lemma I.4 the two factors on the left–hand side are coprime. By unique factorization in $\mathbb{Z}[i]$ we may then write for some unit $\eta$ and $\alpha, \beta \in \mathbb{Z}[i]$:

$$2 + iY = \eta\alpha^3 \ , \ 2 - iY = \eta^{-1}\beta^3.$$

Any unit of $\mathbb{Z}[i]$ is also a cube, so we may assume that $\eta = 1$. Putting $\alpha = m + in$ we must have $\beta = m - in$, and so

$$2 + iY = (m + in)^3 \ , \ 2 - iY = (m - in)^3. \qquad \text{(I.2)}$$

Expanding out the right–hand sides, adding the equations and dividing by 2, we find that

$$m(m^2 - 3n^2) = 2.$$

Hence $m = \pm 1$ or $\pm 2$, and we deduce that the only possibilities are $(m, n) = (-1, \pm 1)$ and $(2, \pm 1)$. We then deduce from equation (I.2), and the original equation, that the only solution in this case is $(X,Y) = (5, \pm 11)$.

$Y$ **even.** In this case $X$ is also even, so we may write $X = 2T$ and $Y = 2Z$. Our equation then becomes

$$Z^2 + 1 = 2T^3.$$

This equation factorizes in $\mathbb{Z}[i]$ as

$$(1 + iZ)(1 - iZ) = (1 + i)(1 - i)T^3.$$

Now $1 \pm i$ has norm 2, and hence these two elements are irreducible. So by unique factorization $1+i$ divides one of the factors on the left–hand side, and by Lemma I.5, the highest common factor of $1 + iZ$ and $1 - iZ$ is $1 + i$.

By unique factorization we may then write

$$1 + iZ = (1 + i)(m + in)^3 \; , \; 1 - iZ = (1 - i)(m - in)^3.$$

This leads to the equation

$$(m + n)(m^2 - 4mn + n^2) = 1.$$

Hence the only solutions are given by $(m, n) = (1, 0)$ or $(0, 1)$. Substituting back we find $Z = \pm 1$ and $(X, Y) = (2, \pm 2)$. □

You should note that we really did use the unique factorization property of the ring $\mathbb{Z}[i]$. The failure of unique factorization in other examples is really the least of our worries. It is not clear at first sight how much of the above proof constitutes a general methodology and how much uses special techniques which are only applicable to this example.

We factorized the equation over the field $\mathbb{Q}(i)$; however, we could have used $\mathbb{Q}(\sqrt[3]{4})$. This may have produced other problems or it may have led to an easier proof. What would be the best strategy for a general curve of the form $Y^2 = X^3 + d$? Should we use $\mathbb{Q}(\sqrt{d})$ or $\mathbb{Q}(\sqrt[3]{-d})$?

We reduced the finding of the integral points on the elliptic curve to the determination of the solutions to equations such as

$$(m + n)(m^2 - 4mn + n^2) = 1$$

and

$$m(m^2 - 3n^2) = 2.$$

Some obvious questions come to mind: Can we always reduce the problem to determining solutions to $F(m, n) = r$, where $F(m, n)$ is a cubic form? Are such cubic forms always reducible? What do we do if we cannot easily spot all the solutions to the equations $F(m, n) = r$, as we did in this example?

We shall answer these questions in later chapters, by giving a general method which works for all elliptic curves. For the rest of this book we shall concentrate on trying to develop general methods which answer the questions above, not only for elliptic curves, but for various other classes of diophantine equations. We would like methods which we can teach a computer to perform. Therefore the methods should not require any insight or experience of a human mathematician to guide them.

# Part 1

# Basic solution techniques

CHAPTER II

# Local methods

In this chapter we give give a brief overview of $p$-adic numbers and various methods from what could be called '$p$-adic numerical analysis'. This will be very sketchy as the material is covered in many textbooks such as those by Cassels [**24**] and Koblitz [**108**].

## II.1. $p$-adic numbers

Let $p$ denote a prime number. Every non-zero rational number, $r$, can be written uniquely in the form

$$r = p^{\alpha} n/d,$$

where $\alpha, n, d \in \mathbb{Z}$, $d \geq 1$ and $(p, n) = (p, d) = (n, d) = 1$. The $p$-adic absolute value of $r$ is defined to be

$$|r|_p = p^{-\alpha}.$$

For completeness we define $|0|_p = 0$ and $\mathrm{ord}_p(r) = \alpha$.

The $p$-adic absolute value gives a multiplicative homomorphism from $\mathbb{Q}^*$ to $\mathbb{R}^*$ which satisfies the triangle inequality

$$|a + b|_p \leq |a|_p + |b|_p.$$

As such it behaves rather like the ordinary absolute value that we all know and love. However, the $p$-adic metric is rather different in that it satisfies the stronger ultra-metric inequality

$$|a + b|_p \leq \max\{|a|_p, |b|_p\}.$$

Two absolute values (or valuations) are defined to be equivalent if they induce the same topology, i.e. they define essentially the same metric. It turns out that all non-trivial metrics on the rational numbers are equivalent either to the standard metric or to a $p$-adic metric for some prime number $p$. For the rest of this book we shall ignore the existence of the trivial metric on $\mathbb{Q}$. The standard metric on $\mathbb{Q}$ will sometimes be called the 'infinite metric' or the metric corresponding to the 'infinite prime'.

Usually one completes the rationals, $\mathbb{Q}$, to form the reals, $\mathbb{R}$, using the standard absolute value by forming the set of all Cauchy sequences with the same limit, convergence being measured in the sense of the standard absolute value. Using the $p$-adic absolute value the same construction can be carried out. But now instead of ending up with the field of real numbers we end up with the field of $p$-adic numbers, $\mathbb{Q}_p$.

We can think of a $p$-adic number as a formal base $p$ expansion which encodes properties modulo higher and higher powers of $p$. Every non-zero $p$-adic number, $n$, can be written in the form

$$n = p^\alpha \left( \sum_{i=0}^{\infty} n_i p^i \right),$$

where $n_i \in \{0, \ldots, p-1\}$ and $(p, n_0) = 1$, we define $\operatorname{ord}_p(n) = \alpha$ and $|n|_p = p^{-\alpha}$. The $p$-adic integers, $\mathbb{Z}_p$, are defined to be the elements $n \in \mathbb{Q}_p$ with $\alpha = \operatorname{ord}_p(n) \geq 0$ (equivalently $|n|_p = p^{-\alpha} \leq 1$). It is clear that $\mathbb{Z}_p$ contains a copy of $\mathbb{N}$, as elements of $\mathbb{N}$ are just the elements of $\mathbb{Z}_p$ for which all but finitely many of the coefficients, $n_i$, are zero.

Obviously one cannot hold a $p$-adic number to infinite precision within a computer's memory, just as one cannot hold a real number to infinite precision. It is usual to work to a given accuracy, so we hold a $p$-adic number as a triple, $(\alpha, \beta, \gamma)$, where $\alpha, \beta \in \mathbb{Z}$, $\gamma \in \mathbb{Z} \cup \{\infty\}$ with $\alpha \leq \gamma$ and $0 \leq \beta < p^{\gamma-\alpha}$, where either $\beta = 0$ or $\gcd(p, \beta) = 1$. Such a triple gives a representation of a $p$-adic number $n$ up to the $p^\gamma$ digit:

$$n = p^\alpha \beta + O(p^\gamma).$$

It is then an easy matter to define the basic operations on $p$-adic numbers in a completely elementary and algorithmic way:

**Addition and subtraction.** Let $z = x \pm y$; then we have

$$\begin{aligned}
\gamma_z &= \min(\gamma_x, \gamma_y), \\
\alpha_z &= \operatorname{ord}_p \left( p^{\alpha_x} \beta_x \pm p^{\alpha_y} \beta_y \pmod{p^{\gamma_z}} \right), \\
\beta_z &= p^{-\alpha_z} \left( p^{\alpha_x} \beta_x \pm p^{\alpha_y} \beta_y \pmod{p^{\gamma_z}} \right).
\end{aligned}$$

**Multiplication.** Let $z = xy$; then we have

$$\begin{aligned}
\alpha_z &= \alpha_x + \alpha_y, \\
\gamma_z &= \min\left(\alpha_x + \gamma_y, \alpha_y + \gamma_x\right), \\
\beta_z &= \beta_x \beta_y \pmod{p^{\gamma_z - \alpha_z}}.
\end{aligned}$$

**Division.** Let $z = x/y$. We can only compute this when $\beta_y \neq 0$, otherwise we obtain an undefined object (rather like division by zero in standard real numbers). Note that if $\beta_y = 0$, we may not be dividing by zero but by something we cannot recognize as different from zero. So assuming $\beta_y \neq 0$, we have

$$\begin{aligned}
\alpha_z &= \alpha_x - \alpha_y, \\
\gamma_z &= \min\left(\gamma_x - \alpha_y, \alpha_x + \gamma_y - 2\alpha_y\right), \\
\beta_z &= \beta_x \beta_y^{-1} \pmod{p^{\gamma_z - \alpha_z}}.
\end{aligned}$$

As examples of this arithmetic of $p$-adic numbers let us consider the 3-adic numbers

$$\sigma = 3 + 2 \cdot 3^2 + 3^4 + O(3^5) \ , \ \tau = 2 \cdot 3^{-1} + 3^2 + 2 \cdot 3^3 + O(3^4).$$

So in this case we have that

$$\begin{array}{lll} \alpha_\sigma = 1, & \beta_\sigma = 34, & \gamma_\sigma = 5, \\ \alpha_\tau = -1, & \beta_\tau = 191, & \gamma_\tau = 4. \end{array}$$

We then have that

$$\begin{aligned} \sigma + \tau &= 2 \cdot 3^{-1} + 3 + O(3^4), \\ \sigma \cdot \tau &= 2 + 3 + 3^2 + O(3^4), \\ \sigma / \tau &= 2 \cdot 3^2 + 2 \cdot 3^3 + 3^4 + 2 \cdot 3^5 + O(3^6) \end{aligned}$$

Just as one forms algebraic number fields by polynomial extension of the rationals, so one can form finite extensions of the $p$-adic numbers by forming polynomial extensions:

$$\mathbb{Q}_p[X]/(f\mathbb{Q}_p[X])$$

where $f \in \mathbb{Q}_p[X]$ is irreducible. The problem of computing in such finite extensions is solved in just the same way as when using extensions of the rationals, i.e. by using polynomial arithmetic but this time with coefficients in $\mathbb{Q}_p$ represented by the triplets discussed above.

The process of taking such finite extensions must eventually terminate in the algebraic closure of $\mathbb{Q}_p$ denoted $\mathbb{Q}_p^a$. However, $\mathbb{Q}_p^a$ is not complete, in the sense that there exist Cauchy sequences in $\mathbb{Q}_p^a$ which have no limit. The completion of $\mathbb{Q}_p^a$, denoted $\Omega_p$, is complete and algebraically closed. The $p$-adic valuation on $\mathbb{Q}_p$ given by $|x|_p = p^{-\alpha_x}$ extends essentially uniquely to each extension field.

We can also arrive at finite extensions of $\mathbb{Q}_p$ in another way which we now recap. Let $K$ denote an algebraic number field. Each prime ideal, $\mathfrak{p}$, defines a valuation on the field $K$ which gives rise to a completion. The valuation is given by looking at the prime ideal factorization of the principal ideal generated by the element $\phi \in K$. If

$$(\phi) = \mathfrak{p}^\alpha \mathfrak{a}/\mathfrak{b},$$

where $\mathfrak{a}, \mathfrak{b}$ are integral ideals and $\mathfrak{p}$ is coprime to $\mathfrak{a}$ and $\mathfrak{b}$, then

$$|\phi|_\mathfrak{p} = p^{-\alpha f_\mathfrak{p}},$$

where $f_\mathfrak{p}$ denotes the residue degree of the ideal $\mathfrak{p}$. In what follows we shall use $e_\mathfrak{p}$ to denote the ramification index of $\mathfrak{p}$ and $p$ is the rational prime lying below $\mathfrak{p}$. Such a completion will be a finite extension of $\mathbb{Q}_p$ which contains $K$, which we shall denote by $K_\mathfrak{p}$. In such a way one obtains an embedding of the number field into $\Omega_p$. This is similar to the usual $s + t$ embeddings of $K$ into $\mathbb{C}$, where $s$ denotes the number of real embeddings and $t$ the number of complex conjugate embeddings. Because of the similarity we shall often refer to these $s + t$ embeddings into the complex numbers as giving $s + t$ 'infinite' valuations and denote them by $|.|_\infty$. It then turns out that all inequivalent valuations on $K$ are in a one to one correspondence with the set of $s+t$ infinite valuations and the valuations arising from each prime ideal.

There are then the following correspondences for a number field $K=\mathbb{Q}(\theta)$ defined by a monic irreducible polynomial $f \in \mathbb{Z}[X]$ such that $f(\theta)=0$. We let $p$ denote a prime (which could include $\infty$).

Let $f=f_1\cdots f_r$ denote the factorization of $f$ into irreducible factors in $\mathbb{Q}_p$. Then each non-conjugate embedding into $\Omega_p$ (or $\mathbb{C}$) corresponds to one of the $f_i$. Two embeddings are said to be conjugate if they map $\theta$ onto a root of the same $f_i$. Such conjugate embeddings give the same valuation on $K$.

If $p \neq \infty$ each factor $f_i$ corresponds to a prime ideal $\mathfrak{p}$ lying above $p$. In such a situation if $\sigma$ is the corresponding embedding into $\Omega_p$ then we have, for any $\phi \in K$, the following identities:

$$\begin{aligned} |\phi|_{\mathfrak{p}} &= |\sigma(\phi)|_p^{e_{\mathfrak{p}} f_{\mathfrak{p}}} = p^{-f_{\mathfrak{p}} e_{\mathfrak{p}} \mathrm{ord}_{\mathfrak{p}}(\phi)} \\ &= N_{K/\mathbb{Q}}(\mathfrak{p})^{-\mathrm{ord}_{\mathfrak{p}}(\phi)} \\ &= |N_{K_{\mathfrak{p}}/\mathbb{Q}_p}(\phi)|_p. \end{aligned}$$

The completion of $K$ with respect to a prime ideal, $\mathfrak{p}$, which we shall denote by $K_{\mathfrak{p}}$, is called a 'local field'. The degree of the polynomial $f_i$, and hence the degree of the extension of $\mathbb{Q}_p$, is given by $e_{\mathfrak{p}} f_{\mathfrak{p}}$. The residue field, $k$, is defined to be the quotient field $K_{\mathfrak{p}}/(\mathfrak{p})$. The residue field is a finite field of degree $f_{\mathfrak{p}}$ over $\mathbb{F}_p$.

One of the most important elementary facts in number theory is the following generalization of Fermat's little theorem,

THEOREM II.1 (Fermat's little theorem). *Let $K_{\mathfrak{p}}$ denote a local field and let $\alpha \in K_{\mathfrak{p}}$. If $\mathrm{ord}_{\mathfrak{p}}(\alpha)=0$ then*

$$\mathrm{ord}_{\mathfrak{p}}(\alpha^{N_{K_{\mathfrak{p}}/\mathbb{Q}}(\mathfrak{p})-1}-1)>0.$$

PROOF. This follows from the fact that the number of elements in the residue field is equal to $N_{K/\mathbb{Q}}(\mathfrak{p})=p^{f_{\mathfrak{p}}}$. □

Another way of stating this theorem is that

$$\alpha^{N_{K_{\mathfrak{p}}/\mathbb{Q}}(\mathfrak{p})-1} \equiv 1 \pmod{\mathfrak{p}}.$$

An element, $\alpha$, of a local field which satisfies $\mathrm{ord}_{\mathfrak{p}}(\alpha)=0$ is called a unit, as it is an integer of the local field whose multiplicative inverse is also an integer. Fermat's little theorem tells us that any unit can be made congruent to one modulo $\mathfrak{p}$ by just raising it to some power which is a divisor of $N_{K_{\mathfrak{p}}/\mathbb{Q}}(\mathfrak{p})-1$.

For example, consider the field $K=\mathbb{Q}(\sqrt{-1})$ and the element $\alpha=1+\sqrt{-1}$. As $\alpha$ has norm 2 it is a unit of $K_{\mathfrak{p}}$ for any prime ideal $\mathfrak{p}$ not lying above 2. The ideal $\mathfrak{p}=(3)$ is prime in $K$ and so has residue degree 2. Fermat's little theorem says that if we raise $\alpha$ to the power of some divisor of $3^2-1=8$

then we obtain an element which is congruent to one modulo 3.

$$\begin{aligned}(1+\sqrt{-1})^2 &= 2\sqrt{-1},\\ (1+\sqrt{-1})^4 &= -4 = 2+3+2\cdot 3^2+2\cdot 3^3+\cdots,\\ (1+\sqrt{-1})^8 &= 16 = 1+2\cdot 3+3^2.\end{aligned}$$

Hence we need to raise $\alpha$ to the eighth power to achieve the desired result.

We let $M_K$ denote the set of all inequivalent valuations on the field $K$. The normalization we have chosen is so that the 'product formula' holds, i.e.

$$\prod_{v\in M_K} |\phi|_v = 1.$$

To see this let $\phi \in K$ and divide the valuations up into two sets: one set being $M_K^\infty$ which consists of the infinite valuations and the other set being $M_K^0$ which consists of the valuations arising from the prime ideals of $K$. Then $M_K = M_K^\infty \cup M_K^0$ and we notice that

$$\prod_{v\in M_K^\infty} |\phi|_v = |N_{K/\mathbb{Q}}(\phi)|$$

and

$$\prod_{v\in M_K^0} |\phi|_v = \prod_{\mathfrak{p}} N_{K/\mathbb{Q}}(\mathfrak{p})^{-\mathrm{ord}_{\mathfrak{p}}(\phi)} = |N_{K/\mathbb{Q}}(\phi)|^{-1},$$

as $|N_{K/\mathbb{Q}}(\phi)|$ is equal to the norm of the ideal generated by $\phi$.

Analogous to the $p$-adic integers $\mathbb{Z}_p$ we can define the local integers in $K_{\mathfrak{p}}$: these are the elements $\alpha \in K_{\mathfrak{p}}$ which satisfy $|\alpha|_{\mathfrak{p}} \le 1$. This gives a rather neat way of defining the ring of integers of a number field $K$ as it is the subring of $K$ which satisfies

$$\mathcal{O}_K = \{\alpha \in K : |\alpha|_{\mathfrak{p}} \le 1 \text{ for all prime ideals } \mathfrak{p}\}.$$

Later on we shall need to discuss $S$-integers and $S$-units. It is convenient to define them now. If $S$ denotes a finite set of inequivalent valuations on a number field $K$, including all the infinite ones, we define the $S$-integers by

$$\mathcal{O}_S = \{\alpha \in K : |\alpha|_v \le 1 \text{ for all valuations } v \notin S\}$$

and the $S$-units by

$$\mathcal{O}_S^* = \{\alpha \in K : |\alpha|_v = 1 \text{ for all valuations } v \notin S\}.$$

The $S$-units of a number field are, by the unit theorem of Dirichlet and Chevalley [**113**, Chapter V], a finitely generated abelian group. The rank of the $S$-units is equal to the number of elements in $S$ minus one.

For example, let $K = \mathbb{Q}$ and $S = \{2, 3, \infty\}$. Then the $S$-integers of $K$ are the numbers of the form

$$\{c/2^a 3^b : c \in \mathbb{Z}, a, b \in \mathbb{N}\};$$

they clearly form a subring of $K$. The $S$-units are the group

$$\{\pm 2^a 3^b : a, b \in \mathbb{Z}\};$$

they clearly form a finitely generated abelian group of rank 2.

One of the most important sets of functions we shall need to consider is height functions. Height functions bring together all the local information about a point in projective space and then give a measure as to how 'big' the point is.

We firstly consider projective $n$-space over $K$, $\mathbb{P}^n_K$. On this space we define a 'local' height function for every $v \in M_K$ by

$$\lambda_{v,K} : \left\{ \begin{array}{ccc} \mathbb{P}^n_K & \longrightarrow & \mathbb{R} \\ (x_1, \dots, x_n) & \longmapsto & \max(|x_1|_v, \dots, |x_n|_v) \end{array} \right.$$

The global height is then given by the formula

$$H_K(x_1, \dots, x_n) = \prod_{v \in M_K} \lambda_{v,K}(x_1, \dots, x_n).$$

However, this definition is dependent on which field we are considering the $x_i$ to lie in. Hence we define the following absolute height:

$$H(x_1, \dots, x_n) = H_K(x_1, \dots, x_n)^{1/[K:\mathbb{Q}]}.$$

Such an absolute height does not depend on the choice of the field $K$ within which we think of the $x_i$ lying. Mostly we shall work with the logarithmic absolute height, or Weil height, given by:

$$h(x_1, \dots, x_n) = \log H(x_1, \dots, x_n).$$

So, to summarize, we have a height function called the logarithmic absolute height (or just height) given by the formula

$$h(x_1, \dots, x_n) = \frac{1}{[K:\mathbb{Q}]} \sum_{v \in M_K} \max(\log |x_1|_v, \dots, \log |x_n|_v).$$

For any element $\alpha \in K$ we define $h(\alpha)$ to mean the height of the projective point $(1, \alpha) \in \mathbb{P}^1_K$. Just to confuse the issue the literature mentions at least two other measures (or heights) of elements, $\alpha$, in a number field. If $\alpha$ has minimal polynomial $f(X)$ of degree $d$ and leading coefficient $a_0$ then one often sees mention of the functions $H_0(\alpha)$ and $M(\alpha)$. $H_0(\alpha)$ is the 'height' of the minimal polynomial, which is the maximum of the absolute values of the coefficients of $f(X)$, while $M(\alpha)$ is the Mahler height given by

$$M(\alpha) = a_0 \prod_{i=1}^{d} \max(1, |\alpha_i|)$$

where the product is over all the roots, $\alpha_i$, of $f(X)$. We have the following relationships:

$$h(\alpha) = \frac{\log M(\alpha)}{d}, \quad M(\alpha) \le \sqrt{d+1} H_0(\alpha), \quad h(\alpha) \le \frac{1}{d} \left( \log H_0(\alpha) + \log d \right).$$

The height function, $h(\alpha)$, is closely linked with the arithmetic of $K$ via the inequalities

$$h(\alpha\beta) \le h(\alpha) + h(\beta)$$

$$h(\alpha^n) = |n|h(\alpha) \text{ if } 0 \neq n \in \mathbb{Z},$$
$$h(\alpha_1 + \cdots + \alpha_n) \leq h(\alpha_1) + \cdots + h(\alpha_n) + \log n.$$

To really see what the logarithmic height is telling you it is perhaps best to look at the simplest example: let $r = n/d \in \mathbb{Q}$ denote a rational number in lowest terms. It is then easy to see that

$$h(r) = \max\{\log|n|, \log|d|\}.$$

It is then clear that there are only finitely many rational numbers with bounded height. Such a theorem is also true in general.

THEOREM II.2. *For any fixed number field $K$ and positive integer $n$ there are only finitely many elements of $\mathbb{P}^n_K$ with absolute logarithmic height bounded by any given constant.*

For more on height functions we refer you to the books by S. Lang [**112**] or Silverman [**172**] which have excellent accounts of the basic theory.

## II.2. $p$-adic numerical analysis

In this section we discuss a few topics which one sometimes meets in a numerical analysis course in the context of the real numbers, namely issues of finding roots of polynomials to arbitrary precision using Newton's formula, computing solutions to power series equations to arbitrary accuracy and providing algorithms to compute transcendental functions to a given accuracy. The analogues of these problems could all be considered to come from an area of '$p$-adic numerical analysis'.

**II.2.1. Newton–Raphson.** Suppose we are given a monic polynomial $f(X) \in \mathbb{Z}_p[X]$ and we wish to compute a root of this polynomial in $\mathbb{Z}_p$. One obvious way of doing this would be to mimic the Newton–Raphson method that is used in the real case. This method is so successful and important that it is named after the person who first used it in the $p$-adic context (namely Hensel). Hensel's lemma plays a fundamental role in many algorithms in computer algebra such as polynomial factorization. What Hensel's lemma does is to provide a criterion for when a solution modulo $p^n$ can be made into a solution modulo $p^{n+1}$. We say the solution modulo $p^n$ is 'lifted' to a solution modulo $p^{n+1}$. This process can then be repeated to lift the solution modulo $p^{n+1}$ to a solution modulo $p^{n+2}$ and so on.

THEOREM II.3 (Hensel's lemma). *Let $f(X) \in \mathbb{Z}_p[X]$ be monic and let $a_0 \in \mathbb{Z}_p$ denote an approximation to the value of a root of $f(X)$ such that*

$$|f(a_0)|_p \leq p^{-2\delta-1},$$

*where $\delta = \mathrm{ord}_p(f'(a_0))$. Then the following sequence tends to a root $a \in \mathbb{Z}_p$:*

$$a_{n+1} = a_n - \frac{f(a_n)}{f'(a_n)}.$$

*In addition the limit, $a$, is the unique root of $f(X)$ satisfying*

$$|a - a_0|_p < p^{-\delta}.$$

We break the proof up into stages. Firstly we prove the following lemma

LEMMA II.4. *We have for all $n \in \mathbb{N}$,*

$$\begin{aligned} |f(a_n)|_p &\leq p^{-2\delta-n-1}, \\ |a_n - a_{n-1}|_p &\leq p^{-\delta-n}. \end{aligned}$$

PROOF. We prove this by induction assuming the result holds for all values less than or equal to $N$. By the second assumption there is a $b \in \mathbb{Z}_p$ such that

$$a_N = a_{N-1} + p^{\delta+N} b.$$

So then

$$f'(a_N) = f'(a_{N-1} + p^{\delta+N} b) = f'(a_{N-1}) + O(p^{\delta+N}).$$

But then

$$\mathrm{ord}_p(f'(a_N)) = \mathrm{ord}_p(f'(a_{N-1})) = \cdots = \mathrm{ord}_p(f'(a_0)) = \delta,$$

hence our first assumption implies that

$$\left|\frac{f(a_N)}{f'(a_N)}\right|_p \leq p^{-\delta-N-1}.$$

Thus

$$|a_{N+1} - a_N|_p \leq p^{-\delta-(N+1)},$$

which proves our second assertion. To prove the first assertion we need to apply Taylor's theorem,

$$f(a_{N+1}) = f(a_N) - f'(a_N)\left(\frac{f(a_N)}{f'(a_N)}\right) + \left(\frac{f(a_N)}{f'(a_N)}\right)^2 c = \left(\frac{f(a_N)}{f'(a_N)}\right)^2 c,$$

where $c \in \mathbb{Z}_p$. Hence we find

$$|f(a_{N+1})|_p \leq p^{-2\delta-2(N+1)} \leq p^{-2\delta-(N+1)-1}.$$

The initial case of $N = 1$ is trivial, so we have proved the lemma. □

PROOF. [Of Hensel's lemma] Using the previous lemma it is clear that the sequence converges to a zero of the polynomial $f(X)$. Hence we have only to show that this is the unique zero within the required range. Suppose that there is another root $\alpha$ such that

$$|\alpha - a|_p \leq p^{-\delta-1}.$$

We shall show that $|\alpha - a_N|_p \leq p^{-\delta-N-1}$ implies that $|\alpha - a_{N+1}|_p \leq p^{-\delta-N-2}$, from which the result will follow. Again using Taylor's theorem we find that (putting $p^{\delta+N+1}b = \alpha - a_N$ for some $b \in \mathbb{Z}_p$) there is a $c \in \mathbb{Z}_p$ such that

$$f(a_N) + f'(a_N)p^{\delta+N+1}b + p^{2\delta+2N+2}b^2 c = f(\alpha) = 0.$$

Hence we obtain

$$p^{\delta+N+1}b = -\frac{f(a_N)}{f'(a_N)} + O(p^{\delta+2N+1}),$$

and so

$$\alpha = a_{N+1} + O(p^{\delta+N+2}).$$

□

For example, let $p$ denote an odd prime and consider the polynomial $f(X) = X^2+1$. Clearly a solution of this equation modulo $p$ can be considered as an element, $\alpha_0$, of $\mathbb{Z}_p$ such that

$$f'(\alpha_0) = 2\alpha_0 \not\equiv 0 \pmod{p}.$$

Hence by Hensel's lemma we can 'lift' a solution modulo $p$ to a solution in $\mathbb{Z}_p$. E.g. $X^2+1$ has the following solution in $\mathbb{Z}_5$:

$$2 + 1\cdot 5 + 2\cdot 5^2 + 1\cdot 5^3 + 3\cdot 5^4 + 4\cdot 5^5 + 2\cdot 5^6 + 3\cdot 5^7 + \cdots$$

So Hensel's lemma provides a mechanism to lift an approximate solution modulo an appropriate power of $p$ to a unique solution in $\mathbb{Z}_p$.

**II.2.2. Power series in one variable.** Let $a_n$ be a sequence of $p$-adic numbers; then the series $\sum a_i$ converges when $a_n \to 0$ (in the $p$-adic sense). This gives a rather nice convergence criterion for power series. Let $f(X) = a_0 + a_1X + a_2X^2 + \cdots$ denote a power series with $p$-adic coefficients. Then this converges at a point $x$ if and only if $a_ix^i \to 0$. Hence it will converge for all values of $x$ if

$$\limsup |a_i|_p^{1/i} = 0,$$

i.e. the $a_i$ become very highly divisible by $p$ as $i$ increases.

The main result we shall require on power series in one variable is the following theorem due to Strassmann, which allows us to bound the number of zeroes of such a series in the $p$-adic numbers.

THEOREM II.5 (Strassmann). *Let $a_i$ denote a sequence of $p$-adic numbers, not all zero, and let*

$$f(X) = \sum_{i\geq 0} a_iX^i$$

*denote a power series which converges for all $x \in \mathbb{Z}_p$, i.e. $|a_i|_p \to 0$. Define $N$ such that*

$$\begin{aligned} |a_N|_p &= \max |a_i|_p, \\ |a_i|_p &< |a_N|_p \text{ for all } i > N. \end{aligned}$$

*Then there are at most $N$ elements $\alpha \in \mathbb{Z}_p$ such that*

$$f(\alpha) = 0.$$

PROOF. Once again we use induction. Firstly we prove the initial step and suppose $N = 0$ and that there actually is an $\alpha \in \mathbb{Z}_p$ such that $f(\alpha) = 0$. Hence

$$\begin{aligned} |a_0|_p &\leq |\sum_{i\geq 1} a_i\alpha^i|_p, \text{ as } \alpha \text{ is a root,} \\ &\leq \max_{i\geq 1} |a_i|_p, \\ &< |a_0|_p, \text{ as } N = 0, \end{aligned}$$

which is a contradiction.

We now prove the induction step and assume that $N > 0$ and that the theorem is true for $N - 1$. Let $\alpha$ denote a fixed zero of $f(X)$. If no such $\alpha$ exists then we are done. We define a new function $g(X)$ by

$$g(X) = \sum b_i X^i$$

where

$$b_i = \sum_{j\geq 0} a_{i+1+j}\alpha^j.$$

We then find that:

1.

$$|b_i|_p \leq \max_{j\geq 0} |a_{i+1+j}|_p \leq |a_N|_p.$$

2.

$$|b_{N-1}|_p \leq \max_{i\geq N} |a_i|_p.$$

Then as $\alpha \in \mathbb{Z}_p^*$ and $N \geq 0$ we find that $|b_{N-1}|_p = |a_N|_p$.

3. If $i \geq N$ we find that

$$|b_i|_p \leq \max_{j>N} |a_j|_p < |a_N|_p.$$

Hence we see that the power series $g(X)$ satisfies the conditions of the theorem but for $N - 1$. By our inductive hypothesis there are then at most $N - 1$ elements $\beta \in \mathbb{Z}_p$ such that $g(\beta) = 0$. We finally have to show that this implies that $f(X) = 0$ has at most $N$ solutions. We already know the existence of one, namely $\alpha$. But

$$\begin{aligned} f(X) &= f(X) - f(\alpha) = \sum_{i\geq 1} a_i(X^i - \alpha^i), \\ &= (X - \alpha)g(X). \end{aligned}$$

Hence any solution of $f(X) = 0$ is either a solution of $g(X) = 0$ or equal to $\alpha$. So there are at most $N$ solutions to $f(X) = 0$. □

**II.2.3. Many power series in many variables.** We shall assume we are given $n$ power series in $n$ variables with coefficients coming from $\mathbb{Z}_p$. We let $\vec{f}$ denote such a vector of power series. We define the Jacobian matrix of such a system by

$$Jac_{\vec{f}}(\vec{x}) = (\partial f_i/\partial x_j).$$

The determinant of the Jacobian matrix we shall denote by $J_{\vec{f}}(\vec{x})$.

We shall require the following standard result on formal power series.

LEMMA II.6. *Let $\vec{f}$ denote an $n$-vector of power series in $n$ variables with no constant term. Suppose $J_{\vec{f}}(\vec{0}) \in \mathbb{Z}_p^*$. Then $\vec{f}$ has an 'inverse' vector of power series with respect to composition of functions.*

PROOF. See [**83**]. □

This result is used to prove

THEOREM II.7 (multi-dimensional Hensel). *We again let $\vec{f}$ denote an $n$ vector of power series in $n$ variables. Suppose there is a vector $\vec{a} \in \mathbb{Z}_p^n$ such that*

$$\begin{aligned} \vec{f}(\vec{a}) &\equiv \vec{0} \pmod{p^{2\delta+1}}, \\ \text{where } \delta &= \operatorname{ord}_p(J_{\vec{f}}(\vec{a})) < \infty. \end{aligned}$$

*Then there is a unique zero of the system of power series $\vec{\alpha}$ such that*

$$\vec{\alpha} \equiv \vec{a} \pmod{p^{\delta+1}}.$$

This is completely analogous to the standard multi-dimensional version of the Newton–Raphson algorithm in ordinary numerical analysis.

PROOF. Just as in the proof of Hensel's lemma we prove this using a Taylor expansion,

$$\vec{f}(\vec{a} + p^\delta \vec{X}) = \vec{f}(\vec{a}) + Jac_{\vec{f}}(\vec{a}) p^\delta \vec{X} + p^{2\delta} \vec{r}(\vec{X}).$$

The remainder power series $\vec{r}(\vec{X})$ will have zero constant and first degree terms. We then define the new vector of power series

$$\vec{g}(\vec{X}) = \vec{X} + A\vec{r}(\vec{X}),$$

where $A$ is the unique matrix such that

$$A Jac_{\vec{f}}(\vec{a}) = p^\delta I_n.$$

The vector of power series $\vec{g}(\vec{X})$ has an inverse, by Lemma II.6, with respect to composition of functions $\vec{g}^{-1}$; this inverse also has no constant terms. We then find

$$\begin{aligned} \vec{f}(\vec{a} + p^\delta \vec{g}^{-1}(\vec{X})) &= \vec{f}(\vec{a}) + Jac_{\vec{f}}(\vec{a}) p^\delta \vec{g}\left(\vec{g}^{-1}(\vec{X})\right), \\ &= \vec{f}(\vec{a}) + Jac_{\vec{f}}(\vec{a}) p^\delta \vec{X}, \end{aligned}$$

We know that $\vec{f}(\vec{a}) = p^{2\delta}\vec{b}$, where $\vec{b}$ is a vector congruent to $\vec{0}$ modulo $p$. We then define

$$\vec{\alpha} = \vec{a} + p^{\delta}\vec{g}^{-1}(-A\vec{b}).$$

Then

$$\begin{aligned} \vec{f}(\alpha) &= \vec{f}(\vec{a}) - AJac_{\vec{f}}(\vec{a})p^{\delta}\vec{b}, \\ &= \vec{f}(\vec{a}) - p^{2\delta}\vec{b} = 0. \end{aligned}$$

That $\alpha$ is the unique such vector follows from the fact that the matrix $A$ has determinant equal to a unit in $\mathbb{Z}_p$. Hence $\vec{x} = -A\vec{b}$ is the unique solution to the equation

$$p^{2\delta}\vec{b} + Jac_{\vec{f}}(\vec{a})p^{\delta}\vec{x} = 0,$$

and $\vec{x}$ is congruent to $\vec{0}$ modulo $p$ as $\vec{b}$ is. □

**II.2.4. The Iwasawa logarithm.** While we are talking about analogues of results and problems in standard numerical analysis we shall discuss how to compute $p$-adic logarithms. Firstly we look at the usual Taylor series expansion of the normal real logarithm about the point 1,

$$\log(1+x) = \sum_{i\geq 1} \frac{(-1)^{i+1}x^i}{i},$$

which satisfies the identity

$$\log((1+x)(1+y)) = \log(1+x) + \log(1+y).$$

We could define a $p$-adic logarithm by taking the above series as a definition. However, we have to worry about convergence problems.

Now if $z \in \Omega_p$ and if $|z-1|_p < 1$ we define the $p$-adic logarithm by the same series

$$\log_p(z) = -\sum_{i\geq 1} \frac{(1-z)^i}{i},$$

which certainly converges. In such a region of convergence we therefore also have the identity

$$\log_p((1+x)(1+y)) = \log_p(1+x) + \log_p(1+y).$$

In the region where $|z|_p < p^{-1/(p-1)}$ we also have that

$$\mathrm{ord}_p\left(\log_p(1+z)\right) = \mathrm{ord}_p z.$$

We would like to define a logarithm for the whole of $\Omega_p$. We do this using an idea of Iwasawa with the following rules:

- For all $x, y \in \Omega_p$ we have $\log_p(xy) = \log_p(x) + \log_p(y)$.
- If $\omega$ is a root of unity in $\Omega_p$ and $s \in \mathbb{Z}$ then $\log_p(\omega p^s) = 0$.

Using the above definition we can evaluate the $p$-adic logarithm at any point $\alpha \in \Omega_p$.

In our later examples $\alpha$ will be a unit of some $K_\mathfrak{p}$ where $K$ is some number field and $\mathfrak{p}$ is a prime ideal. So we shall assume that this case holds for convenience. Note that $\alpha \in K_\mathfrak{p}$ implies that $\log_p(\alpha) \in K_\mathfrak{p}$ as $K_\mathfrak{p}$ is complete. We let $e$ denote the ramification index of $\mathfrak{p}$ and $f$ the residue degree. By Fermat's little theorem we know that the order of the image of $\alpha$ in the residue field $\mathbb{F}_{p^f}$ divides $p^f - 1$. We can hence compute the order of the image of $\alpha$ in $\mathbb{F}_{p^f}$, call it $o$. This can be done by using either a naive method or the Baby-Step-Giant-Step method, see [**32**]. For elements of large finite fields the determination of $o$ may not be that easy, however in the examples which interest us the field will be relatively small.

Now note that if we choose $t$ such that $p^t > e$, and assume $p$ is odd, then

$$(1-\alpha^o)^{p^t} = 1 - p^t\alpha^o + \cdots - \alpha^{op^t}$$

and so $\operatorname{ord}_\mathfrak{p}(1-\alpha^{op^t}) > \operatorname{ord}_\mathfrak{p}(\mathfrak{p}^{p^t}) > 1$. It is easily verified that the last inequality also holds for $p = 2$. Then

$$\log_p(\alpha) = \frac{1}{op^t}\log_p(\alpha^{op^t}) = \frac{-1}{op^t}\sum_{i\geq 1}\frac{(1-\alpha^{op^t})^i}{i}.$$

We are hence left only with the task of studying how fast such a series converges and developing techniques to speed the convergence up. We shall want to know how many terms to take to obtain a desired level of accuracy, a question which is answered by the following result:

LEMMA II.8. *Let* $\operatorname{ord}_p(1-z) \geq 1$ *and let* $M$ *denote an arbitrary given integer. We let* $N$ *denote the smallest integer solution of*

$$n \geq \frac{1}{\operatorname{ord}_p(1-z)}\left(\frac{\log n}{\log p} + M\right).$$

*Then we have*

$$\log_p z = -\sum_{i=1}^{N}\frac{(1-z)^i}{i} + O(p^M).$$

PROOF. First note that $\operatorname{ord}_p n \leq \log n/\log p$ for all positive integers $n$. Now, if $n \geq N$, we have

$$\begin{aligned}
\operatorname{ord}_p\left(\frac{(1-z)^n}{n}\right) &= n\operatorname{ord}_p(1-z) - \operatorname{ord}_p n \\
&\geq \left(\frac{\log n}{\log p} + M\right) - \frac{\log n}{\log p} \\
&\geq M.
\end{aligned}$$

Hence

$$\mathrm{ord}_p\left(-\sum_{i\geq N}\frac{(1-z)^i}{i}\right)\geq M.$$

From which the desired result follows. □

---

**Algorithm for $p$-adic logarithms**

---

DESCRIPTION: Finds the $p$-adic logarithm of the algebraic number $\alpha \in K$ with respect to the embedding of $K$ into $\Omega_p$ given by the ideal $\mathfrak{p}$.
$\alpha$ is assumed to be a unit of $K_{\mathfrak{p}}$

INPUT: $\alpha \in K$, a prime ideal, $\mathfrak{p}$, of $\mathcal{O}_K$ and a natural number $M$.

OUTPUT: The $p$-adic logarithm $\beta$ up to an accuracy of $p^M$.

1. Compute $o$ such that $\mathrm{ord}_{\mathfrak{p}}(\alpha^o - 1) > 0$.
2. Set $\gamma = \alpha^{op^t}$ where $t$ is chosen to be the smallest number such that $m = \mathrm{ord}_p(\gamma - 1) \geq \frac{1}{2}\mathrm{ord}_p(D(\theta)) + 1$.
3. Compute the smallest integer solution, $n$, to $n \geq \left(\frac{\log n}{\log p} + M\right)/m$.
4. Set $\beta := 0$ and $\delta := 1 - \gamma$.
5. For $i = 1, \ldots, n$ do
6. $\quad\beta := \beta - \delta/i$.
7. $\quad\delta := \delta(1 - \gamma)$.
8. Enddo.
9. $\beta := \beta/(op^t)$.

---

In such an algorithm we need to take care of any coefficient swell. If $K = \mathbb{Q}(\theta)$ we can write $\gamma - 1$ as a polynomial in $\theta$. We can assume that no coefficient has a denominator divisible by $p$, hence we can assume that $\gamma - 1 \in \mathbb{Z}_p[\theta]$. By the choice of $o$ and $t$ the polynomials representing $\beta$ and $\delta$ have no coefficients with $p$-adic value greater than one. For the reason for the choice of $t$ see the proof of Lemma VI.4. Hence we may reduce every coefficient in the algorithm by taking its value modulo

$$p^{M+\log M/\log p}.$$

This allows us to take care of the possible coefficient swell.

Suppose one wanted to take the 3-adic logarithm of the rational integer 2. First we need to compute an exponent $o$ such that $2^o \equiv 1 \pmod 3$. Clearly we can take $o = 2$, in which case we have

$$\log_3(2) = \frac{\log_3(4)}{2}.$$

Hence we need to compute $\log_3(4)$, but as $4 \equiv 1 \pmod 3$ this can be done from the series

$$\begin{aligned}\log_3(4) &= -\sum_{i\geq 1}\frac{(1-4)^i}{i},\\ &= -\left(-3+\frac{9}{2}-9+\frac{81}{4}-\frac{243}{5}+\frac{243}{2}+O(3^7)\right),\\ &= 3+2\cdot 3^2+3^3+2\cdot 3^5+2\cdot 3^6+O(3^7).\end{aligned}$$

Hence

$$\log_3(2) = 2\cdot 3+2\cdot 3^2+3^5+3^6+O(3^7).$$

One way of speeding up the computation of $p$-adic logarithms is to use an observation of de Weger [**208**]. Instead of using the series

$$\log_p z = -\sum_{i\geq 1}\frac{(1-z)^i}{i}$$

we could use instead the series

$$\log_p\left(\frac{1+z}{1-z}\right) = 2\left(\sum_{i\geq 0}\frac{z^{2i+1}}{2i+1}\right) = 2\left(z+\frac{z^3}{3}+\frac{z^5}{5}+\cdots\right).$$

Of course if we make $z$ very close to zero $p$-adically then the above series will converge much faster.

As in the example above, suppose we want to compute $\log_3(2)$. Again this is easy once we have computed $\log_3(4)$. We find

$$\begin{aligned}\log_3(4) &= \log_3\left(\frac{1+\frac{3}{5}}{1-\frac{3}{5}}\right)\\ &= 2\left(\frac{3}{5}+\frac{9}{125}+\frac{243}{15625}+\cdots\right)\\ &= 3+2\cdot 3^2+3^3+2\cdot 3^5+2\cdot 3^6+O(3^7).\end{aligned}$$

Of course this section would not be complete without a discussion of the $p$-adic exponential function. This is defined by

$$\exp_p z = \sum_{i\geq 0}\frac{z^n}{n!},$$

which converges if $\mathrm{ord}_p z > 1/(p-1)$. It also satisfies the following formulae, in the region in which it is defined:

$$\begin{aligned}(1+z)^a &= \exp_p\left(a\log_p(1+z)\right),\\ \exp_p(z_1+z_2) &= \exp_p(z_1)\exp_p(z_2),\\ \mathrm{ord}_p z &= \mathrm{ord}_p\left(\exp_p(z)-1\right).\end{aligned}$$

Finally we notice that we have

LEMMA II.9. *Let $\alpha \in \Omega_p$ denote a p-adic unit, If*

$$\mathrm{ord}_p(\alpha - 1) > 1/(p-1)$$

*then*

$$\mathrm{ord}_p(\alpha - 1) = \mathrm{ord}_p(\log_p \alpha).$$

PROOF. This follows from the above equalities for $\exp_p$. □

## II.3. Exercises

1). Let $\vec{f}$ denote an $n$-vector of power series in $n$ variables with no constant term. Show that if $J_{\vec{f}}(\vec{0}) \in \mathbb{Z}_p^*$ then $\vec{f}$ has an 'inverse' vector of power series with respect to composition of functions.

2). Show that there are only finitely many $\vec{x} \in \mathbb{P}_K^n$ for a number field $K$ with bounded height.

3). Determine the $p$-adic roots of the following polynomials up to the tenth $p$-adic digit for $p = 2, 3, 5, 7$:

$$\begin{array}{ll} X^2+2, & X^3+3X-1, \\ X^4+1, & X^4+2X-1. \end{array}$$

4). Compute the 5-adic logarithms of the following rational numbers

$$3/5\ ,\ 5/3\ ,\ 16\ ,\ 3\ ,\ 1/3.$$

5). Compute the 3-adic logarithms of the following algebraic numbers

$$1+\sqrt{-3},\ 1+\sqrt{-1},\ 1-(-1)^{3/4}.$$

6). Let $\Omega$ denote a 7-adic root of the polynomial

$$X^2 + 4 + 5\cdot 7 + 4\cdot 7^2 + 5\cdot 7^4 + O(7^5).$$

Determine all the solutions, up to an accuracy of $O(7^3)$, to the simultaneous equations

$$\begin{aligned} 5s + t + 6\Omega t + 7(s + t + \Omega) + 7^2(t^2 + \Omega - 1) + O(7^3) &= 0, \\ 5s + t + \Omega t + 7(2t + \Omega) + 7^2(s + t + \Omega - 1) + O(7^3) &= 0. \end{aligned}$$

CHAPTER III

# Applications of local methods to diophantine equations

In this chapter we give some so–called 'local' considerations which either allow us to completely solve a diophantine equation, aid us in locating the solutions or give us information about the solutions which can be used in a more advanced method. We show how to apply the $p$-adic analysis of the last chapter to find solutions to equations using Skolem's method and then finally we discuss how various pieces of local information can be put together in an algorithmic manner using sieving. Sieving is no more than a catch–all phrase for a process meaning applying local considerations one after another to sieve out (or remove) non-solutions. The idea behind sieving is that anything left after we have used a sieve has a good chance of being an actual solution.

## III.1. Applications of Strassmann's theorem

We shall now give three examples where we can apply Strassmann's theorem, Theorem II.5, to deduce information about diophantine equations. In all three cases we derive a $p$-adic power series and then apply Strassmann's theorem to bound the number of solutions to the diophantine equation. Its range of application is, however, rather limited.

**III.1.1.** $X^3 + 6Y^3 = \pm 1$**.** We shall now show that the equation

$$X^3 + 6Y^3 = \pm 1,$$

where we are only interested in solutions with $(X, Y) \in \mathbb{Z}^2$, has only the trivial solutions $(X, Y) = (\pm 1, 0)$.

Firstly consider the algebraic number field $K = \mathbb{Q}(\theta)$, where $\theta^3 + 6 = 0$. Why, you may ask, choose this number field? This should be the one which springs immediately to mind in such a situation as we can write our equation as

$$N_{K/\mathbb{Q}}(X - \theta Y) = \pm 1.$$

The field $K$ is a cubic number field with one real embedding, it therefore has a single fundamental unit which is given by $1 + 6\theta + 3\theta^2$. Such a fundamental unit can be determined quite easily using the modern methods explained in [**32**] and [**34**], or using a computer package to perform the calculation for you. It is clear that the only units of finite order in $K$ are $\pm 1$.

By considering the factorization of our Thue equation

$$(X - \theta Y)(X - \theta\omega Y)(X - \theta\omega^2 Y) = \pm 1,$$

where $\omega$ is a non-trivial cube root of unity, we see from the unique factorization of the ideal $(X - \theta Y)\mathcal{O}_K$ that we must have

$$X - \theta Y = \pm(1 + 6\theta + 3\theta^2)^k.$$

We can then formally expand the right hand side as a power series in $k$, to obtain

$$X - \theta Y = \pm\left(1 + 3(\theta^2 k + 2\theta k) + 9(2\theta^2(k^2 + k)) + 27(\ldots)\right). \tag{III.1}$$

We then notice that $X^3 + 6$ is irreducible over $\mathbb{Q}_3$, a fact which we can deduce either by actually trying to produce a non-trivial factorization or by noticing that there is only one prime ideal of $K$ lying above the ideal (3). We can hence equate coefficients of $\theta$ in equation (III.1). The coefficient of $\theta^2$ then gives us

$$0 = \pm(3k + 9(\ldots)).$$

From Strassmann's theorem we then deduce that there is only one 3-adic solution to the above 3-adic power series. But we already know one solution, namely $k = 0$, which corresponds to our known solutions of the original equation. Hence $(X, Y) = (\pm 1, 0)$ are the only solutions.

**III.1.2.** $X^3 + 2Y^3 = \pm 1$**.** We now consider the Thue equation

$$X^3 + 2Y^3 = \pm 1.$$

This only has the integral solutions $(X, Y) = \pm(1, 0)$ and $\pm(1, -1)$, as we shall now show.

We consider the field $K = \mathbb{Q}(\theta)$ where $\theta^3 + 2 = 0$. In this field we again have one fundamental unit, namely $-1-\theta$. Consideration of the factorization of $X^3 + 2Y^3$ leads us to the equation

$$X - \theta Y = \pm(-1 - \theta)^k.$$

Naively applying the method above will not give us any $p$-adic power series to which to apply Strassmann's theorem. What worked in the first example was that the fundamental unit was congruent to 1 modulo 3 and hence the power series in $k$ which we obtained converged 3-adically.

By Fermat's little theorem we know that for every algebraic integer, $\alpha$, of $K$ and every coprime prime ideal, $\mathfrak{p}$, we have $\alpha^o \equiv 1 \pmod{\mathfrak{p}}$, where $o$ divides $p^{f_\mathfrak{p}} - 1$. By raising $\alpha^o$ to the $p^t$ where $p^t > e_\mathfrak{p}$ we obtain (as we did in our previous discussion of $p$-adic logarithms)

$$\alpha^{op^t} \equiv 1 \pmod{p}.$$

In our example if we consider the prime ideal lying above (3), which completely ramifies, we see that

$$(-1 - \theta)^3 = 1 - 3\theta(1 + \theta).$$

Hence we should consider the following three equations:

$$X - \theta Y = \begin{cases} \pm(1 - 3\theta(1+\theta))^s & \text{when } k = 3s, \\ \pm(1+\theta)(1 - 3\theta(1+\theta))^s & \text{when } k = 1 + 3s, \\ \pm(1+\theta)^2(1 - 3\theta(1+\theta))^s & \text{when } k = 2 + 3s. \end{cases}$$

We expand the right hand side of these equations as power series in $s$ and then equate coefficients of $\theta^2$ as before to obtain three 3-adic power series in $s$ which have to be zero for a solution to our original diophantine equation. These three power series are then given by

$$0 = \begin{cases} 6s + 9(\dots) & \text{when } k = 3s, \\ 6s + 9(\dots) & \text{when } k = 1 + 3s, \\ 1 + 9(\dots) & \text{when } k = 2 + 3s. \end{cases}$$

We deduce there is at most one solution, $s$, to the first two 3-adic power series equations and there is no solution to the third equation. By inspection we see that our original equation has a solution when $k = 0$ and $k = 1$. Hence these two solutions must be the only solutions. So the only solutions are given by $(X, Y) = \pm(1, 0)$ and $\pm(1, -1)$.

**III.1.3.** $X^3 + 6XY^2 - Y^3 = \pm 1$. In this example Strassmann's theorem will also show us where to look for a solution as well. We shall show that the only solutions to the Thue equation

$$X^3 + 6XY^2 - Y^3 = \pm 1$$

are given by $(X, Y) = \pm(1, 0), \pm(0, 1), \pm(1, 6)$.

To see this consider the field $K = \mathbb{Q}(\theta)$ where $\theta^3 + 6\theta - 1$. In $K$ there is one fundamental unit given by $\theta$. We notice that

$$\theta^3 = (1 - 6\theta) \equiv 1 \pmod 3$$

and that there is only one ramified prime ideal lying above 3. We look at the three 3-adic power series, given by setting $a = 0, 1$ or 2 in the following equation :

$$\begin{aligned} X - \theta Y &= \theta^k = \theta^a (1 - 6\theta)^s, \\ &= \theta^a \left(1 - 6\theta s + 18\theta^2 s^2 + 27(\dots)\right), \end{aligned}$$

from which we deduce that there are at most six solutions, two when $k \equiv 1 \pmod 3$ and four when $k \equiv 0 \pmod 3$. We easily find the solutions $k = 0, 1$ which correspond to $(X, Y) = \pm(1, 0), \pm(0, 1)$. The other two solutions must lie in the family $k \equiv 0 \pmod 3$ which suggests we look at $k = \pm 3, \pm 6, \dots$. Luckily we find the final two solutions at $k = 3$.

The above example shows how we can use $p$-adic arguments to locate solutions as well as bound the number of actual solutions. From these examples it appears that the method works for all examples of cubic Thue equations of negative discriminant. This is however rather optimistic. We leave it as an exercise to construct an example where the arguments above do not work. For example try one of the examples above but with a different prime.

It also appears from the above examples that we need to use primes for which there is only one prime ideal lying above it. This is not true but using such primes makes the presentation neater. For more general primes one needs to decide on which prime ideal to choose and then find a $p$-adic power series which must be zero for a solution to exist. We cannot just equate coefficients of $\theta^2$ in the general case. We can however find a suitable $p$-adic power series by, for instance, using Siegel's identity (see the next section).

## III.2. Skolem's method

In the last section we saw how, if we could produce a $p$-adic power series in one variable, we could bound the number of solutions to a diophantine equation. However, we would have to be dealing with very small problems for the above method to work all the time. An obvious extension would be to generalize the method to the case when we obtain power series in many variables. In such a situation we will require many power series as well. The idea behind this solution method, often called Skolem's method, is to generalize Hensel's lemma rather than Strassmann's theorem. Then after a finite amount of 'sieving' we can hopefully locate all the solutions. In any case we will at least obtain an upper bound on the number of solutions if this method works.

The method dates back to Skolem and his school in the 1930s. Until the 1980's it was the main method used to solve many diophantine equations. Its popularity has since waned with the advent of the 'magic' LLL–algorithm which we shall come to later. However, we shall see later that the modern methods and Skolem's method often share the sieving process in common. The sieving process will turn out to be the major bottleneck. Hence from a computational point of view Skolem's method, when it works, is often no worse than the modern methods. We shall explain this method with an example.

**III.2.1.** $X^4 - 2Y^4 = \pm 1$. We shall now show that the Thue equation,

$$X^4 - 2Y^4 = \pm 1,$$

has at most 12 solutions in integers. To study this equation we first have to consider the quartic number field $K = \mathbb{Q}(\theta)$, where $\theta^4 - 2 = 0$. The unit rank of the ring of integers is two and we can take as a pair of fundamental units the elements

$$\eta_1 = 1 + \theta^2 \ , \ \eta_2 = 1 + \theta.$$

We therefore have to determine all possible pairs $a_1, a_2$ to the equation

$$X - \theta Y = \beta = \pm \eta_1^{a_1} \eta_2^{a_2}.$$

The smallest prime number which stays prime in $K$ is 5 and in the residue field the image of $\eta_1$ has order 12 and the image of $\eta_2$ has order 312, indeed

$$\eta_1^{12} = 1 + 5 \cdot 2\theta^2 + 5^2(\dots)\ ,\ \eta_2^{312} = 1 + 5(4\theta^2 + 3\theta^3) + 5^2(\dots).$$

Hence we could equate coefficients of $\theta^2$ and $\theta^3$ in the identity

$$X - \theta Y = \pm\eta_1^{b_1}\eta_2^{b_2}(1 + (\eta_1^{12} - 1))^{k_1}(1 + (\eta_2^{312} - 1))^{k_2}$$

to find two power series in the two variables $k_1$ and $k_2$. However, we would have to do this for all possible values of the $b_i$ which range $0 \leq b_1 \leq 11$, $0 \leq b_2 \leq 311$. Hence this looks rather an unpromising situation.

We instead notice that over the algebraic closure of $\mathbb{Q}$ we have four equations of the form

$$X - \theta_i Y = \beta_i$$

which correspond to the four roots of our polynomial $X^4 - 2$. Eliminating $X$ and $Y$ from these four equations gives us two equations for the $\beta_i$, namely

$$(\theta_i - \theta_2)\beta_1 + (\theta_1 - \theta_i)\beta_2 + (\theta_2 - \theta_1)\beta_i = 0 \qquad \text{(III.2)}$$

for $i = 3, 4$. This last equation is often referred to as Siegel's identity. Now the prime 7 decomposes in the field $K$ as a product of three prime ideals, one of degree 2 and two of degree 1. In other words, modulo 7 the polynomial $x^4 - 2$ factorizes as a product of two linear and one quadratic polynomial:

$$x^4 - 2 \equiv (x + 2)(x + 5)(x^2 + 4) \pmod 7,$$

as 7 is not an index divisor. We take $\theta_1, \theta_2$ to be the 7-adic roots of $x^4 - 2$ given by

$$\theta_1 = 2 + 7(\dots)\ ,\ \theta_2 = 5 + 7(\dots).$$

We then take $\theta_3 = \Omega$ and $\theta_4 = \Omega'$ to be the roots of the 7-adic quadratic polynomial

$$g(x) = x^2 + 4 + 5 \cdot 7 + 4 \cdot 7^2 + 5 \cdot 7^4 + O(7^5) = (x^4 - 2)/(x - \theta_1)(x - \theta_2).$$

In the two degree–one 7-adic localizations of $K$ the elements $\eta_i$ both satisfy

$$\eta_i^6 \equiv 1 \pmod 7.$$

In the quadratic 7-adic localization of $K$ we find that the $\eta_i$ satisfy

$$\eta_1^6 \equiv 1 \pmod 7\ ,\ \eta_2^{48} \equiv 1 \pmod 7.$$

We write $a_1 = b_1 + 6k_1$ and $a_2 = b_2 + 48k_2$. We first need to determine which values of $0 \leq b_1 \leq 5$ and $0 \leq b_2 \leq 47$ solve the following congruences, which come from Siegel's identity:

$$\begin{aligned}(\theta_i - \theta_2)\eta_1^{(1)b_1}\eta_2^{(1)b_2} \quad &+(\theta_1 - \theta_i)\eta_1^{(2)b_1}\eta_2^{(2)b_2} \\ &+(\theta_2 - \theta_1)\eta_1^{(i)b_1}\eta_2^{(i)b_2} \equiv 0 \pmod 7\end{aligned}$$

for $i = 3, 4$. To do this we need only loop through the $6 \times 48$ possibilities for the $b_i$ and test these in the previous equation. We find that there are 6 possible pairs $(b_1, b_2)$ given by

$$(b_1, b_2) = (0,0), (0,1), (2,23), (3,24), (3,25), (5,47).$$

Then given these solutions we need to expand the equations in (III.2) as two 7-adic power series in the variables $k_1$, $k_2$. We obtain the following 7-adic power series, $f_1$ and $f_2$, in each of our six cases:

1. $b_1 = b_2 = 0$.

$$\begin{aligned} f_1 &= 5k_1 + k_2 + 6\Omega k_2 + 7(\dots), \\ f_2 &= 5k_1 + k_2 + \Omega k_2 + 7(\dots). \end{aligned}$$

2. $b_1 = 0$, $b_2 = 1$.

$$\begin{aligned} f_1 &= 5k_1 + 6k_2 + 5\Omega k_1 + 7(\dots), \\ f_2 &= 5k_1 + 6k_2 + 2\Omega k_1 + 7(\dots). \end{aligned}$$

3. $b_1 = 2$, $b_2 = 23$.

$$\begin{aligned} f_1 &= 4 + 5k_1 + 3k_2 + \Omega(2k_1 + 5k_2) + 7(\dots), \\ f_2 &= 4 + 5k_1 + 3k_2 + \Omega(5k_1 + 2k_2) + 7(\dots). \end{aligned}$$

4. $b_1 = 3$, $b_2 = 24$.

$$\begin{aligned} f_1 &= 4 + 2k_1 + 6k_2 + \Omega(4 + k_2) + 7(\dots), \\ f_2 &= 4 + 2k_1 + 6k_2 + \Omega(3 + 6k_2) + 7(\dots). \end{aligned}$$

5. $b_1 = 3$, $b_2 = 25$.

$$\begin{aligned} f_1 &= 5 + 2k_1 + k_2 + \Omega(1 + 2k_1) + 7(\dots), \\ f_2 &= 5 + 2k_1 + k_2 + \Omega(6 + 5k_1) + 7(\dots). \end{aligned}$$

6. $b_1 = 5$, $b_2 = 47$.

$$\begin{aligned} f_1 &= 6 + 2k_1 + 4k_2 + \Omega(5k_1 + 2k_2) + 7(\dots), \\ f_2 &= 6 + 2k_1 + 4k_2 + \Omega(2k_1 + 5k_2) + 7(\dots). \end{aligned}$$

In each of the above six cases we apply Theorem II.7 to find that in each case there is exactly one possible solution in $\mathbb{Q}_7^2$. As every one of these cases corresponds to two solutions of our Thue equation, $X^4 - 2Y^4 = \pm 1$, we have an upper bound on the number of solutions of 12. A trivial search reveals 6 solutions:

| $b_1$ | $b_2$ | X | Y |
|---|---|---|---|
| 0 | 0 | −1 | 0 |
| 0 | 0 | 1 | 0 |
| 0 | 1 | 1 | −1 |
| 0 | 1 | −1 | 1 |
| 5 | 47 | −1 | −1 |
| 5 | 47 | 1 | 1 |

Hence there could exist another six possible solutions. To eliminate, or find, these one could either use another prime or apply some of the methods in later chapters. However, our method has at least told us that the remaining solutions (if they exist at all) lie in one of the three families

$$(i)\ a_1 \equiv 2 \pmod 6,\ a_2 \equiv 23 \pmod{48},$$

$$(ii)\ a_1 \equiv 3 \pmod 6,\ a_2 \equiv 24 \pmod{48},$$

$$(iii) a_1 \equiv 3 \pmod 6,\ a_2 \equiv 25 \pmod{48}.$$

We also know that each family can only contain at most one pair of solutions. This idea of finding congruence conditions on the exponents of identities satisfied by solutions of diophantine equations will come up again when we discuss sieving an $S$-unit equation. We can treat the first part of the above method as 'sieving' out the six possible families of $(b_1, b_2)$ from the 288 possible families. Hence although Skolem's method has not worked, using the prime 7, it has given us information which we could use later in the more advanced methods.

## III.3. The Hasse principle

We sometimes use local considerations to show that no solutions exist to certain diophantine equations. Every local field (be it $\mathbb{Q}_p, \mathbb{R}$ or $K_\mathfrak{p}$) contains a copy of $\mathbb{Q}$. Hence if a solution to a given equation exists then there is a solution in every such local completion. This often gives us a very easy check as to when a diophantine equation is not soluble.

As a rather silly example consider the equation

$$x^2 + y^2 = -1.$$

This equation certainly has no real solutions. It hence has no integral (or rational) solutions either. As another example consider the equation

$$x^2 - 17y^2 = 7.$$

This has no solutions in $\mathbb{Q}_7$ as the congruence $x^2 \equiv 3 \pmod 7$ has no solutions. Therefore this equation also has no rational solutions either.

The previous two examples are special cases of the projective plane curve of degree 2, i.e. a ternary quadratic form:

$$aX^2 + bY^2 + cZ^2 = 0.$$

Given $a, b, c \in \mathbb{Z}$ we are interested in determining whether such an equation has a solution in relatively prime integers $(X, Y, Z)$ which are not simultaneously zero. Obviously we need first check whether it has a solution in $\mathbb{Q}_p$ for every prime number $p$. We shall show how to do this in Chapter IV. It turns out that this is all we need do

THEOREM III.1 (Hasse). *The equation*

$$aX^2 + bY^2 + cZ^2 = 0$$

*has a non-trivial solution in* $\mathbb{Z}^3$ *if and only if it has a non-trivial solution in* $\mathbb{Q}_p^3$ *for every prime* $p$ *(including* $\infty$*).*

PROOF. An exercise, which can be found in many books which discuss quadratic forms such as [**25**, Chapter 5]. □

You should note that this theorem gives us a little more than is really required. It can be shown that the number of primes (including $\infty$) for which the equation $aX^2 + bY^2 + cZ^2 = 0$ is not locally soluble is always even.

Given the above example we have the following definition.

A diophantine equation is said to satisfy the **Hasse principle** if the existence of rational (global) solutions is guaranteed by the existence of $p$-adic (local) solutions for every prime $p$ (including $\infty$).

The Hasse principle is also often called the local–global principle. We see that the equations of Hasse's theorem satisfy the Hasse principle. However, we are not always so lucky. The standard example of an equation which does not satisfy the Hasse principle is

$$3X^3 + 4Y^3 + 5Z^3 = 0$$

due to Selmer. This has no rational solutions but has local solutions in every $p$-adic field. We shall return to the failure of the Hasse principle later when we discuss elliptic curves.

## III.4. Finding small solutions

Sometimes when one solves a diophantine equation one has a bound on the solution space or one is only interested in 'small' solutions. It would be nice if there was a fast method to locate all solutions up to any given bound. In terms of the language of Section II.1 we wish to determine all the solutions with bounded logarithmic height. We could just run through all possibilities checking each one in turn. We shall call this the naive method. It is easy to see that this naive method applied to an equation in two variables would take at least $O(e^{2B})$ operations, if $B$ were the bound on the logarithmic height.

This naive method is analogous to the naive method for finding prime numbers. The naive method for prime numbers takes each integer in turn and checks whether it is a prime. However, for over 2000 years a much better technique has been known, namely the *Sieve of Eratosthenes.* In this method composite numbers are eliminated (or sieved out) by using small prime numbers which have already been found. Generalizations of the Sieve of Eratosthenes have been used for theoretical purposes in the analytic theory of numbers for many years. They are also used in algorithms for factoring and solving the discrete logarithm problem in groups such as $\mathbb{F}_q^*$. We shall use a computational analogue of this sieving procedure to find the required small solutions.

Like the Sieve of Eratosthenes we use information gleaned from considering small prime numbers to remove large numbers of non-solutions from consideration. In other words we look at where the solutions could be locally, using mod $p$ or $p$-adic arguments. This local information is then put together to deduce information about the location of global solutions. We eliminate as many solutions as possible at the first stage using a single prime. The remaining possible solutions are passed to a second stage where they are checked modulo a different prime $q$ and so on. At each stage one has sieved out a large number of non-solutions.

For example suppose we wish to find all rational solutions to the equation

$$Y^2 = aX^4 + bX^3 + cX^2 + dX + e \tag{III.3}$$

with $h(X) \leq B$, some given positive constant. Firstly we write $X$ as $N/M$ with $N, M \in \mathbb{Z}$ coprime and $M \geq 1$. We then know we must find all solutions to the equation

$$\begin{aligned} (M^2Y)^2 &= F(M,N) \\ &= aN^4 + bMN^3 + cM^2N^2 + dM^3N + eM^4 \ , \ \max(|N|, M) \leq e^B. \end{aligned}$$

Heuristically we believe that for a given, random prime $p$ the expression $F(M,N)$, for any given $M$ and $N$, is a square modulo $p$ about half the time. Hence looking modulo $p$, for a single prime number $p$, will hopefully eliminate half of the solution space.

We define a sieving procedure as follows:

---

**Recursive algorithm for sieving a curve of the form $Y^2 = F(X)$**

---

DESCRIPTION: Sieve(M,N,R):
Finds all solutions to the equation $Y^2 = F(X)$ with $h(X) \leq B$.

INPUT: $M_0, N_0, R \in \mathbb{Z}$.

OUTPUT: Solutions to (III.3) with $h(X) \leq B$ such that $X = N/M$ with $N - N_0 \equiv M - M_0 \equiv 0 \pmod{R}$.

Choose the smallest prime, $p$, such that $\gcd(p, R) = 1$.
For $M_1 = M_0$ to $pR$ step $R$ do,
  For $N_1 = N_0$ to $pR$ step $R$ do,
    If $M_1$ and $N_1$ are not both divisible by $p$ then
      If $F(M_1, N_1)$ is a square modulo $p$ then
        If $pR > 2e^B$ then
          Check if $N_1/M_1$ or $(N_1 - pR)/M_1$ really is a solution and if so print it.

```
                    Else
                         Call Sieve(M_1, N_1, pR).
                    Endif.
               Endif.
          Endif.
     Enddo.
Enddo.
```

This sieving procedure is called via Sieve$(0,0,1)$. It works in a recursive way by assuming we have a solution $(M_0, N_0)$ modulo $R$. It then lifts this solution to a new modulus $pR$ where $p$ is a prime number coprime to $R$. For every solution modulo $pR$ found it calls itself again until the current modulus is greater than $2e^B$. When the current modulus is greater than $2e^B$ the current values are tested to see if they really are global solutions. The method is therefore essentially a *depth first* strategy.

It is probably best for very small primes $p$ to use either prime powers or composite moduli in the loops rather than just the prime $p$ itself. An alternative approach would be to only use primes larger than 5, say. Essentially we have combined local information using a Chinese remainder process to obtain information about possible solutions up to the desired bound. However, here, as each prime taken was coprime to the current modulus, the Chinese remaindering needed was trivial. Later on we shall adapt this method to find an efficient sieving procedure for exponential diophantine equations where the moduli will not be coprime.

It remains to discuss how much faster such a sieving technique will be. We first note that

$$\theta(x) = \sum_{p \le x} \log p \approx O(x),$$

hence the largest prime we need to take is of size $\log e^B = B$, and there are roughly $B/\log B$ primes less than $B$ in size. At each step we eliminate roughly half of the cases modulo $p$; hence the complexity can be estimated by

$$\begin{aligned}
\text{Time} &\approx p_1^2\left(1+\frac{p_2^2}{2}\left(1+\frac{p_3^2}{2}(1+\cdots)\right)\right), \\
&\le B^2\left(1+\frac{B^2}{2}\left(1+\frac{B^2}{2}(1+\cdots)\right)\right), \\
&= \sum_{i=1}^{B/\log B} \frac{B^{2i}}{2^{i-1}}, \\
&= 2B^2\frac{\left((B^2/2)^{B/\log B}-1\right)}{B^2-2} \quad \text{( Sum of a G.P. )} \\
&= O((B^2/2)^{B/\log B}).
\end{aligned}$$

So we see that using sieving gives a slightly better running time than the naive method. Clearly whether sieving is better in practice than the naive method would depend on the implied constants which arise from the implementation. In addition the above analysis of the sieving method has been very pessimistic so as to make the formulae easier to handle.

## III.5. Exercises

1). Determine an upper bound on the number of solutions to the Thue equations

$$x^3 + dy^3 = 1$$

for various values of $d$.

2). Apply the method using Strassmann's theorem to the equation

$$x^3 + 6y^3 = \pm 1$$

for all primes $p \leq 29$.

3). Determine upper bounds on the number of solutions to the Thue equation

$$x^4 + dy^4 = 1$$

for various values of $d$.

4). Find all solutions with $h(x) \leq 7$ to the equation

$$y^2 = x^4 + 1.$$

5). Show that there are exactly six solutions to the equation

$$x^4 - 2y^4 = \pm 1$$

by using Skolem's method for primes other than the one used in the text.

6). Show that

$$3x^3 + 4y^3 + 5z^3 = 0$$

does not satisfy the Hasse principle.

CHAPTER IV

# Ternary quadratic forms

In this chapter we consider the problem of determining the rational solutions to equations of the form

$$Q(x, y, z) = Ax^2 + Bxy + Cxz + Dy^2 + Eyz + Fz^2 = 0,$$

where $A, B, C, D, E, F \in \mathbb{Q}$ and $(x, y, z) \neq (0, 0, 0)$. Clearly determining the integral solutions to such equations is equivalent to determining the rational solutions. We wish to develop procedures to decide whether such an equation has a non-trivial rational solution and if it does have a non-trivial rational solution to then find one in a reasonable amount of time.

We shall show that if such an equation does have a non-trivial rational solution then it has an infinite number of such solutions. This infinite set of solutions can be parameterized, a fact which shall be used later in studying other types of equations. Much of what follows can be found in the book by Mordell [**138**]. We reproduce the methods here as we shall use them in a variety of contexts elsewhere, such as

1. Determining integral points on elliptic curves by reducing to a finite set of Thue equations.
2. Solving discriminant form equations in quartic number fields.
3. Determining a basis for the set of rational points on an elliptic curve.

## IV.1. A normal form

Clearly if we are interested in the rational solutions to $Q(x, y, z) = 0$ we can make a rational change of variables without altering the existence or otherwise of rational solutions. The equation $Q(x, y, z) = 0$ determines a curve of genus 0 in $\mathbb{P}^2$ and we would like to determine a 'better' equation for such a curve which is more convenient to work with.

LEMMA IV.1. *The equation $Q(x, y, z) = 0$ is equivalent to one of the form*

$$ax^2 + by^2 + cz^2 = 0 \qquad \text{(IV.1)}$$

*where $a, b, c \in \mathbb{Z}$ are square free and pairwise coprime.*

PROOF. We first make the change of variable, *if* $A \neq 0$,

$$x \longmapsto x - (By + Cz)/(2A)$$

which eliminates the terms in $xy$ and $xz$. If $A = 0$ then a related change of variable will do the same trick. This puts our curve of genus 0 in the form

$$A'x^2 + D'y^2 + E'yz + F'z^2 = 0.$$

We can then make the similar change of variable

$$y \longmapsto y - E'z/(2D')$$

to eliminate the term in $yz$. Hence we can place our curve in the form

$$ax^2 + by^2 + cz^2 = 0,$$

with $a, b, c \in \mathbb{Q}$. By multiplying through by some common denominator we can assume that $a, b, c \in \mathbb{Z}$.

We now need to show that we can take $a, b, c$ to be square free and pairwise coprime. If $a$ is not square free, then it is of the form $a'a''^2$ and we can make the change of variable $x \mapsto x/a''$ to obtain an equation with a square free value of $a$. Similar considerations will obviously apply to $b$ and $c$. Hence we can suppose that all of $a, b, c$ are square free.

Now suppose that $a, b, c$ are not pairwise coprime, we can however assume that $\gcd(a, b, c) = 1$. Without loss of generality we can assume that $p$ is a prime factor of both $b$ and $c$. i.e. $b = pb'$ and $c = pc'$. Making the change of variable $x \mapsto px$ our equation becomes

$$pax^2 + b'y^2 + c'z^2 = 0.$$

All of the above changes of variables will map rational points to rational points and so the result is proved. □

A ternary quadratic form which satisfies the condition of the above lemma shall hence forward be referred to as reduced. We will remove from any future discussion the case when the curve is degenerate i.e. we shall assume that $abc \neq 0$.

## IV.2. Local solubility

As mentioned in Section III.3, one way of showing the non-existence of rational points is to determine a prime, $p$, for which the equation has no non-trivial solutions in $\mathbb{Q}_p$. We shall assume our curve of genus 0 is reduced, this clearly will not affect the question of whether it has non-trivial solutions in $\mathbb{Q}_p$. So we have

$$Q(x, y, z) = ax^2 + by^2 + cz^2 = 0.$$

For finite places, $p$, we can assume that any non-trivial solution in $\mathbb{Q}_p$ is normalized such that $x, y, z \in \mathbb{Z}_p$ and $\min(\mathrm{ord}_p(x), \mathrm{ord}_p(y), \mathrm{ord}_p(z)) = 0$. We now treat each possibility in turn.

**IV.2.1.** $p = \infty$. Clearly a necessary and sufficient condition for a solution to exist in $\mathbb{R}^3 \setminus \{0,0,0\}$ is that $a, b, c$ are not all of the same sign. By multiplying through by $-1$ and permuting $(x, y, z)$ if necessary we can assume that if $Q(x, y, z) = 0$ has a rational solution then $a, b > 0$ and $c < 0$.

**IV.2.2.** $p$ **odd and** $p$ **does not divide** $abc$. We shall show that in this case we always obtain a $p$-adic solution. First consider the case when the three Legendre symbols

$$\left(\frac{a}{p}\right), \left(\frac{b}{p}\right), \left(\frac{c}{p}\right)$$

are not all equal to each other. We can then, by reordering the variables if necessary, assume that

$$\left(\frac{-b/a}{p}\right) = 1.$$

Hence we have a solution modulo $p$ given by $(x_0, 1, 0)$ where $x_0 \in \mathbb{F}_p^*$. If we then apply Hensel's lemma, Theorem II.3, to the polynomial

$$aX^2 + b = 0$$

then we see that $x_0$ can be lifted to a $p$-adic solution, $x$, as we have $2ax_0 \not\equiv 0 \pmod{p}$.

We are therefore left with the case when all of the three Legendre symbols are equal. In such a situation a solution modulo $p$ cannot exist with any one of $x$, $y$ or $z$ being congruent to zero modulo $p$. We can then assume that $z \equiv 1 \pmod{p}$. Then as $a \not\equiv 0 \pmod{p}$ we have

$$x^2 \equiv a^{-1}(-by^2 - c) \pmod{p}.$$

Taking all possible residue classes for $y$ we see that the right hand side takes a total of $(p+1)/2$ values. So at least one value of $y$ gives a value of the right hand side which is a square modulo $p$, it cannot be zero modulo $p$ by assumption.

To show that $Q(x, y, z) = 0$ is locally soluble in $\mathbb{Q}_p$ we need to show that this solution modulo $p$ can lift to a solution in $\mathbb{Q}_p$. Let our modulo $p$ solution be denoted by $(x_0, y_0, 1)$. Again using Hensel's lemma, Theorem II.3, we can show that there is a $p$-adic solution $(x, y_0, 1)$ with $x \equiv x_0 \pmod{p}$. We can see this by considering the polynomial equation

$$f(x) = ax^2 + by_0^2 + c = 0$$

with $x \in \mathbb{Q}_p$. We already know a solution modulo $p$, namely $x_0$. This will automatically lift by Hensel's lemma to a solution in $\mathbb{Q}_p$ if $\mathrm{ord}_p(f'(x_0)) = 0$, but this is clear as $f'(x) = 2ax$.

**IV.2.3. $p$ odd and $p$ divides $abc$.** By symmetry we can assume that $p$ divides $a$ only. Suppose that we have a solution with $z \equiv 0 \pmod{p}$ then we must also have $y \equiv 0 \pmod{p}$. We can then assume that $x \equiv 1 \pmod{p}$ and we have the equation, with $x = 1 + px_0, y = py_0$ and $z = pz_0$,

$$a(1 + x_0 p)^2 + bp^2 y_0^2 + cp^2 z_0 = 0$$

which is impossible modulo $p^2$ as $a$ is square free. Hence we can assume that $z \not\equiv 0 \pmod{p}$. The equation modulo $p$ then becomes

$$by^2 \equiv -cz^2 \pmod{p},$$

hence if we have a solution then $-bc$ must be a quadratic residue modulo $p$. We can then find a solution modulo $p$ given by $(0, y_0, 1)$ where $y_0^2 \equiv -c/b \pmod{p}$. This will lift to a $p$-adic solution by Hensel's lemma as above on considering the polynomial

$$f(y) = by^2 + c$$

as clearly $y_0 \not\equiv 0 \pmod{p}$.

**IV.2.4. $p$ even.** One can show in this case, although it is rather tedious, that there is a non-trivial solution in $\mathbb{Q}_2$ if and only if there is a non-trivial solution to the congruence

$$ax^2 + by^2 + cz^2 \equiv 0 \pmod{8}.$$

To summarize we have shown:

THEOREM IV.2. *The equation*

$$ax^2 + by^2 + cz^2 = 0$$

*with $a, b, c \in \mathbb{Z}$, pairwise coprime and square free is locally soluble everywhere if and only if*

1. *$a, b, c$ do not have the same sign.*
2. *It is soluble in $\mathbb{Q}_p$ for all primes $p$ which divide $2abc$.*

*In particular this last condition is equivalent to*

1. *There is a non-trivial solution modulo 8.*
2. *$x^2 + bc = 0 \pmod{a}$ and $x^2 + ac = 0 \pmod{b}$ and $x^2 + ab = 0 \pmod{c}$ are also soluble.*

We can eliminate a prime to test here as it is known that the number of primes (including infinity) for which a conic is not soluble is always even. We then have a very fast procedure to determine whether there is a solution everywhere locally. Such a procedure will run in subexponential time as it is only as fast as determining the primes which divide $abc$, and factoring is known to be a subexponential time problem. The problem now is, given we know whether such a solution exists, can we find a solution when we know one does exist?

## IV.3. Global solubility

Suppose we have a curve of genus 0 given by an equation

$$ax^2 + by^2 + cz^2 = 0$$

which we assume to be reduced. We have just shown that we can, in subexponential time, determine whether a solution exists to this equation in $\mathbb{Z}^3 \setminus \{(0,0,0)\}$, as $ax^2 + by^2 + cz^2 = 0$ satisfies the Hasse principle, as was mentioned in Chapter III. We would like to produce a solution: the standard way of doing this is using Holzer's theorem:

THEOREM IV.3 (Holzer). *If there is a non-trivial solution then there is one which satisfies*

$$|x| \leq \sqrt{|bc|}\ ,\ |y| \leq \sqrt{|ca|}\ ,\ |z| \leq \sqrt{|ab|}.$$

PROOF. See [**138**, Page 47] □

The theorem says that if you want to find a non-trivial solution then loop through all the above possibilities. This clearly takes $O(|a|\sqrt{|bc|})$ operations, and is therefore exponential in the length of the input data. This is rather unsatisfactory given that we can decide whether a solution exists in much better than exponential time. It would be nice to be able to obtain such a solution in subexponential time as well.

We shall now give three methods to find such a solution. Each method displays an important principle in solving diophantine equations. However, the first two methods are only meant as illustrative examples; our third method is the one which is the most efficient and which should be used in practice.

**IV.3.1. A sieving method.** Adapting the sieving algorithm of Section III.4 in combination with Holzer's theorem will allow us to find a solution.

---

**Recursive algorithm for sieving a ternary quadratic form**

---

DESCRIPTION: $Sieve(x_0, y_0, M)$:
Finds solutions to the, locally soluble everywhere, equation $ax^2 + by^2 + cz^2 = 0$.

INPUT: $x_0, y_0, M \in \mathbb{Z}$.

OUTPUT: Solutions to (IV.1) with $x, y, z$ bounded by Holzer's theorem with $x - x_0 \equiv y - y_0 \equiv 0 \pmod{M}$.

Choose the smallest prime, $p$, coprime to $M$ and $c$.<br>
For $x_1 = x_0$ to $\max(p\ M, \sqrt{|bc|})$ step $M$ do,<br>
    For $y_1 = y_0$ to $\max(p\ M, \sqrt{|ca|})$ step $M$ do,<br>
        If $p$ does not divide both $x_1$ and $y_1$ then<br>
            If $-(ax_1^2 + by_1^2)/c$ is a square modulo $p$ then<br>
                If $M\ p > \min(\sqrt{|bc|}, \sqrt{|ca|})$ then

```
                    Check if (x_1, y_1, sqrt(-(a x_1^2 + b y_1^2)/c)) is a
                    solution.  If so then print it.
                Else
                    Call Sieve(x_1, y_1, pM).
                Endif.
            Endif.
        Endif.
    Enddo.
Enddo.
```

**IV.3.2. A method using quadratic number fields.** Let $K$ denote a number field, $m \geq 2$ an integer and $S$ a finite set of places of $K$ including all the infinite ones. We let $K(S,m)$ denote a set of representatives of $K^*/K^{*m}$ such that if we adjoin the $m$th root of an element of $K(S,m)$ to $K$ we obtain an extension unramified away from $S$. We shall choose the representatives to be elements of $\mathcal{O}_K$ and to not be divisible by any $m^{th}$-power in $\mathcal{O}_K \setminus \mathcal{O}_K^*$.

Equivalently we have

$$K(S,m) = \{\alpha \in K^*/K^{*m} : \mathrm{ord}_{\mathfrak{p}}(\alpha) \equiv 0 \pmod{m} \text{ for all } \mathfrak{p} \notin S\}.$$

It is clear that $K(S,m)$ is a finite group. If the class number of $K$ is equal to one then it is rather easy to compute as it is nothing but $\mathcal{O}_S^*/\mathcal{O}_S^{*m}$, where $\mathcal{O}_S^*$ denotes the $S$-units of $K$. If the class number is not equal to one then it is a little more tricky to compute, however, if one is willing to assume the generalized Riemann hypothesis (GRH) and other conjectures, it can be computed in subexponential time, as its computation involves only a little linear algebra and the testing of ideals for principality, see [**171**] for the case $m = 2$. The complexity is roughly (assuming GRH etc.)

$$L_{D_K}(1/2, c),$$

for some positive constant $c$, where $D_K$ is the discriminant of the field $K$. The group $K(S,m)$ will occur over and over again as it is used in methods for finding integral points on elliptic curves, solving quartic discriminant form equations and determining the Mordell–Weil groups of elliptic curves and higher genus hyperelliptic curves.

We now return to trying to solve

$$Q(x,y,z) = ax^2 + by^2 + cz^2 = 0$$

for integer pairwise coprime values of $x, y$ and $z$. Clearly this is equivalent to solving

$$x^2 + aby^2 = -acz^2,$$

for integer pairwise coprime values of $x, y$ and $z$. By assumption $ab$ is square free and hence $K = \mathbb{Q}(\sqrt{-ab})$ is a quadratic number field (except in the case

$\{a, b\} = \{-1, 1\}$, which we can ignore as then there is a solution which is easy to spot). We can write the last equation as

$$N_{K/\mathbb{Q}}(x + \sqrt{-ab}y) = -acz^2.$$

Suppose $\mathfrak{p}$ is a prime ideal of $K$ which divides $(x + \sqrt{-ab}y)$ to an odd power, then from the unique factorization of ideals either $\mathfrak{p}$ divides $ac$ or $\mathfrak{p}$ divides $z$. If $\mathfrak{p}$ divides $z$ then it must also divide $x - \sqrt{-ab}y$ and so $\mathfrak{p}$ divides $2x$ and $2\sqrt{-ab}y$. Suppose now that $\mathfrak{p}$ does not divide 2 or $ab$ then it must divide both $x$ and $y$ which contradicts the coprimality of $x, y$ and $z$.

So if we let $S$ denote the set of valuations, $v$, of $K$ for which

$$|2abc|_v \neq 1$$

then, up to multiplication by elements of $K^{*2}$, we have

$$x + \sqrt{-ab}y \in K(S, 2),$$

We are only interested in those elements of $K(S, 2)$ whose norm is equal to $-ac$ modulo $\mathbb{Q}^{*2}$. For such element we can equate coefficients of 1 and $\sqrt{-ab}$ to read off values for $x$ and $y$. Computing a corresponding value of $z$ is then easy. This gives us a conjectured subexponential method of finding a solution to our ternary quadratic form. (I would like to thank S. Siksek for the above observation.) Suppose one wants to solve

$$x^2 + 11y^2 - 3z^2 = 0.$$

We look at this as a norm form over $K = \mathbb{Q}(\sqrt{-11})$,

$$N_{K/\mathbb{Q}}(x + \sqrt{-11}y) = 3z^2.$$

We then compute $K(S, 2)$ for this field $K$, where $S$ is the set of prime ideals dividing $2, 3$ and 11. As $K$ has class number one this really is not necessary but a simple equation will help illustrate the method:

$$\begin{aligned} K(S, 2) &= \langle -1 \rangle \times \langle \sqrt{-11} \rangle \times \langle 2 \rangle \times \langle (1 + \sqrt{-11})/2 \rangle \\ &\quad \times \langle -1 + (1 + \sqrt{-11})/2 \rangle. \end{aligned}$$

We then consider only those elements of $K(S, 2)$ whose norm is equal to 3 modulo squares. There are eight such elements, one of which is $(1+\sqrt{-11})/2$. Hence this tells us that we can take $x = 1/2$ and $y = 1/2$, which leads to $z = 1$. Homogenizing our solution we see that a solution is given by $(1, 1, 2)$.

We clearly do not really need to compute $K(S, 2)$, we only need to compute an element in $K$ whose norm is equal to 3 modulo squares. This is true no matter what the class number is. However, as the group $K(S, 2)$ will come up in later chapters it seemed only right to introduce it at this point.

**IV.3.3. A descent method.** The following method is due to Lagrange and makes use of Fermat's method of descent. The existence of a solution to our desired equation is deduced from the existence of a solution to a 'smaller' equation. The existence of a solution on this 'smaller' equation is in turn deduced from the existence of a solution on an even 'smaller' equation, and so on. Eventually we reach an equation which is so small it is easy to spot a solution.

We start with the equation

$$ax^2 + by^2 + cz^2 = 0$$

with $a, b, c$ being square free and pairwise coprime and which is everywhere locally soluble. We then make the transformation $z \mapsto w/c$ and multiply the equation by $c$ and set $A = ac$ and $B = bc$ to obtain the equation

$$w^2 = Ax^2 + By^2.$$

Clearly $A, B \in \mathbb{Z}$ are square free. We can assume that $|A| \leq |B|$. We shall also assume that $|B| \neq 1$, as if $|B| = 1$ there is a trivial solution.

We now find an $r \in \mathbb{N}$ such that $r^2 \equiv A \pmod{|B|}$, with the choice of the square root $r$ being so that $0 \leq r \leq |B|/2$. That such an $r$ exists is due to our equation being everywhere locally soluble. We then set

$$r^2 - A = BB'd^2 = QB$$

where $B'$ is square free. We then have that

$$|B'| = \left|\frac{r^2 - A}{Bd^2}\right| \leq \left|\frac{r^2 - A}{B}\right| \leq \frac{|B|}{4} + 1 < |B|.$$

If we now make the change of variables

$$\begin{aligned} x &\mapsto \frac{rX - W}{r^2 - A} \\ y &\mapsto \frac{Y}{Bd} \\ w &\mapsto \frac{-AX + rW}{r^2 - A} \end{aligned}$$

then our equation becomes $W^2 = AX^2 + B'Y^2$. All we need now do is repeat the method again and again until we get an equation which has a trivial solution which we can spot.

---

**Descent method for ternary quadrics**

---

| | |
|---|---|
| DESCRIPTION: | *Descent*$(A, B)$: |
| | `Uses Lagrange's method to determine a solution.` |
| INPUT: | `The equation` $w^2 = Ax^2 + By^2$`, which is locally soluble everywhere.` |
| OUTPUT: | `A solution` $(x, y, w)$. |

If $|A| > |B|$ then return $(y, x, w) = Descent(B, A)$.
If $A = 1$ then return $(1, 0, 1)$.
If $B = 1$ then return $(0, 1, 1)$.
Find $r \in [0, \ldots, |B|/2]$ such that $r^2 \equiv A \pmod{B}$.
Let $d2$ be the square part of $Q = (r^2 - A)/B$.
Set $B' = Q/d2$ and $d = \sqrt{d2}$.
Set $(X, Y, W) = Descent(A, B')$.
Put $y = BdY$, $x = rX + W$ and $w = AX + rW$.
Return $(x, y, w)$.

---

It remains to discuss just how fast the above method is. On paper it seems rather fast; however, there are two hidden snags in the method:

- We need to be able to factorize the number $B$ if we are to extract a square root modulo $|B|$.
- We need to be able to factorize $Q$, as the only known way to extract the square part of a number is to completely factorize it.

However, it does give a fast method in practice as we shall now see.

We shall return to our example considered above,

$$x^2 + 11y^2 - 3z^2 = 0$$

which we rewrite as

$$x^2 = 3z^2 - 11y^2.$$

So $A = 3$ and $B = -11$, and we find that if we set $r = 5$ then $r^2 \equiv A \pmod{|B|}$. If we then apply the transformations $x = (-3X + 5Z)/22$, $y = -Y/11$ and $z = (5X - Z)/22$ then we obtain the 'smaller' equation

$$Z^2 = -2Y^2 + 3X^2.$$

We can then apply the method again with $r = 1$, to obtain the 'trivial' equation

$$Z_1^2 = -2X_1^2 + Y_2^2.$$

From which we can find a solution and therefore recover a solution to our original problem.

## IV.4. New solutions for old

Using the methods outlined previously we can determine whether

$$Q(x, y, z) = Ax^2 + Bxy + Cxz + Dy^2 + Eyz + Fz^2 = 0$$

has a non-trivial solution and if so find one. This was achieved by looking at the 'reduced form' $ax^2 + by^2 + cz^2$. Let us now suppose that we have determined a non-trivial rational solution $(x_0, y_0, z_0) \in \mathbb{Z}^3$. We will now attempt to parameterize all other non-trivial solutions in terms of this known one. We will hence obtain new solutions from our known old solution.

By a permutation of the variables we can always assume that $x_0 \neq 0$. We then obtain the general solution using the following formulae:

$$\begin{aligned} x &= rx_0 \\ y &= ry_0 + p \\ z &= rz_0 + q \end{aligned}$$

where $p, q$ and $r$ are rational parameters. We shall assume that $x, y$ and $z$ are also solutions to our quadratic form, with $x, y, z \in \mathbb{Z}$. Hence all solutions to our quadratic form are given by specializing the parameters $p, q$ and $r$. But we cannot just specialize them to anything. It turns out that we can take $p$ and $q$ to take any values we want but that then $r$ will be given if $(x, y, z)$ is to be a solution of $Q(x, y, z) = 0$.

We substitute the above formulae for $x, y$ and $z$ into our quadratic form $Q(x, y, z)$ to deduce

$$r^2 Q(x_0, y_0, z_0) - r(c_1 p + c_2 q) + (c_3 p^2 + c_4 pq + c_5 q^2) = 0$$

where $c_1, \dots, c_5$ are constants which can be easily computed in any given example. Now by assumption $(x_0, y_0, z_0)$ is a known solution and so $Q(x_0, y_0, z_0)$ is equal to zero. Hence we can write

$$r(c_1 p + c_2 q) = (c_3 p^2 + c_4 pq + c_5 q^2).$$

We obtain

$$\begin{aligned} (c_1 p + c_2 q)x &= (c_3 p^2 + c_4 pq + c_5 q^2)x_0, \\ (c_1 p + c_2 q)y &= (c_3 p^2 + c_4 pq + c_5 q^2)y_0 + p(c_1 p + c_2 q), \\ (c_1 p + c_2 q)z &= (c_3 p^2 + c_4 pq + c_5 q^2)z_0 + q(c_1 p + c_2 q). \end{aligned}$$

We now multiply $p$ and $q$ by their common denominator so that they take integer coprime values (we shall also call the resulting new variables $p$ and $q$). Hence for some $g \in \mathbb{Z}$ we have

$$\begin{aligned} gx = f_1(p, q) &= a_{1,1}p^2 + a_{1,2}pq + a_{1,3}q^2, \\ gy = f_2(p, q) &= a_{2,1}p^2 + a_{2,2}pq + a_{2,3}q^2, \\ gz = f_3(p, q) &= a_{3,1}p^2 + a_{3,2}pq + a_{3,3}q^2, \end{aligned} \tag{IV.2}$$

where $a_{i,j} \in \mathbb{Z}$. If we can deduce then that $g$ comes from a finite set of integers then we can express $x, y$ and $z$ as a finite set of quadratic forms in two coprime integer variables $p$ and $q$. Note that $g$ may be equal to zero only in the case when $(c_1 p + c_2 q) = 0$. We shall now assume that $g \neq 0$. If we write $A = (a_{i,j})$ where $a_{i,j}$ are the integers in equation (IV.2) above then we can express (IV.2) as the following matrix equation:

$$g \begin{pmatrix} x \\ y \\ z \end{pmatrix} = A \begin{pmatrix} p^2 \\ pq \\ q^2 \end{pmatrix}.$$

Hence

$$\det(A)\begin{pmatrix} p^2 \\ pq \\ q^2 \end{pmatrix} = g \ \mathrm{adj}(A)\begin{pmatrix} x \\ y \\ z \end{pmatrix}$$

and so $g$ must divide $\det(A)$ as $(p,q)=1$. That $\det(A)$ is always non-zero can be checked by a computer algebra system, indeed in [**71**] it is shown that it is equal to the determinant of the quadratic form. There is then only a finite number of possible values for $g$. There could still be a large choice for the number of such possibilities. In [**71**] the following trick is suggested for reducing this number of possible values. The possible values of $g$ must be such that we can find $p,q \in \mathbb{Z}/g\mathbb{Z}$ which satisfy

$$\begin{aligned} a_{1,1}p^2 + a_{1,2}pq + a_{1,3}q^2 &\equiv 0 \pmod{g}, \\ a_{2,1}p^2 + a_{2,2}pq + a_{2,3}q^2 &\equiv 0 \pmod{g}, \\ a_{3,1}p^2 + a_{3,2}pq + a_{3,3}q^2 &\equiv 0 \pmod{g}, \end{aligned}$$

with $\gcd(p,q,g)=1$.

Continuing the example considered above we wish to express all solutions to

$$x^2 + 11y^2 - 3z^2 = 0$$

in terms of quadratic forms in two integer valued coprime variables $p$ and $q$. We have already found one solution, namely $(1,1,2)$ so we now make the transformation

$$x = r\ ,\ y = r + p\ ,\ z = 2r + q.$$

Substituting these expressions into our quadratic form we find

$$(22p - 12q)r = 11p^2 - 3q^2.$$

Finally scaling $p$ and $q$ to take integer, coprime values we see that the quadratic forms we want are given by

$$\begin{aligned} gx &= 11p^2 - 3q^2, \\ gy &= 33p^2 - 12pq - 3q^2, \\ gz &= 22p^2 + 22pq - 18q^2, \end{aligned}$$

where $g$ is some integer value. If we are interested in rational values for $x, y$ and $z$ then we could assume that $g = 1$ and $p, q \in \mathbb{Q}$ and we would be done. However, our restriction of $p$ and $q$ to coprime integer values and our interest in integral values for $x, y$ and $z$ leads us to the problem of determining $g$.

We first form the matrix $A$,

$$A = \begin{pmatrix} 11 & 0 & -3 \\ 33 & -12 & -3 \\ 22 & 22 & -18 \end{pmatrix}.$$

We wish to determine for which values of $g$ does the following matrix congruence have a solution

$$A \begin{pmatrix} p^2 \\ pq \\ q^2 \end{pmatrix} \equiv \vec{0} \pmod{g}.$$

It is easy to determine all such $g$ by taking the Hermite normal form, see [**155**] or [**32**], of the matrix A. In our case, the (row-reduced) Hermite normal form of $A$ is given by

$$\begin{pmatrix} 11 & 0 & 3 \\ 0 & 2 & 0 \\ 0 & 0 & 6 \end{pmatrix}.$$

Hence we deduce that $g$ must be a divisor of 132, but $g$ must also be such that $6q^2 \equiv 0 \pmod{g}$, $2pq \equiv 0 \pmod{g}$ and $11p^2 + 3q^2 \equiv 0 \pmod{g}$. Now if 4 divides $g$ then 2 must divide both $p$ and $q$, but we have assumed that $p$ and $q$ are coprime. Hence the only possible values for $g$ must be plus or minus a divisor of 66.

A quicker way to see that $g$ must divide 66 is to note that $g$ must divide the largest elementary divisor of $A$. As the Smith normal form of $A$ is $\mathrm{diag}(66, 2, 1)$ we immediately see that $g$ must divide 66.

## IV.5. Exercises

1). Show that $ax^2 + by^2 + cz^2 = 0$ with $a, b, c$ pairwise coprime and square free is solvable in $\mathbb{Q}_2$ if and only if it is solvable modulo 8.

2). Show that the $g$ in Section IV.4 must always divide the largest elementary divisor of $A$.

3). Determine which of the following curves of genus 0 have rational solutions:

(a) $2\,ac + 3\,b^2 + 3\,c^2 - 2\,bc = 0$.
(b) $4\,bc + b^2 + 2\,ab + c^2 = 0$.
(c) $18\,ac + 40\,c^2 + 43\,b^2 + a^2 + 34\,ab + 60\,bc = 0$.
(d) $26\,ac + 35\,c^2 + 53\,b^2 + 6\,a^2 + 40\,ab + 114\,bc = 0$.
(e) $18\,ac + 11\,c^2 + 17\,b^2 + 2\,a^2 - 16\,ab - 30\,bc = 0$.
(f) $4\,ac - 5\,c^2 + b^2 + 2\,a^2 - 6\,ab + 4\,bc = 0$.
(g) $-58\,ac - 138\,c^2 - 93\,b^2 + 15\,a^2 + 18\,ab + 244\,bc = 0$.
(h) $62\,ac + 33\,c^2 + 57\,b^2 + 8\,a^2 - 56\,ab - 94\,bc = 0$.

4). For all of the curves in the above exercise which you have shown do possess rational solutions:

(i) Find a rational solution.
(ii) Parameterize all other solutions in terms of two coprime integer valued variables $p$ and $q$.

5). Show that $ax^2 + by^2 + cz^2 = 0$ satisfies the Hasse principle.

6). Give a method to compute $K(S, m)$ for any number field $K$, any set of places $S$ (which includes all the infinite ones) and any integer $m \geq 2$.

7). For various number fields $K$, sets of primes $S$ and positive integers $m$ compute $K(S, m)$. (This question will almost definitely require the use of some computer system which allows one to compute class groups, units etc.)

CHAPTER V

# Computational diophantine approximation

In this chapter we look at ways of approximating irrational numbers by rationals. We are interested in ways of effectively finding 'good' approximations and ways of showing that one cannot do this too well. By this we mean that the there is a trade off between the accuracy of the rational approximation and the size of the integers that make the rational approximation. In our later applications we will combine this with theoretical diophantine approximation results to produce a practical method to solve a large class of diophantine equations. The results in this chapter provide the computational means by which the theoretical results of Baker and others are converted into practical results.

We start by looking at the one variable case, and then we consider a way of generalizing this to many variables. There are many ways to do this generalization, we present the one which has been the most successful in recent years, namely the algorithm of Lenstra, Lenstra and Lovász. This algorithm is usually referred to as the 'LLL' algorithm in honour of its inventors. It has proved a useful tool in much of computational number theory. For instance we can use it when computing units and class groups of number fields and it can also be used to factor rational polynomials. We shall concentrate on its use in solving diophantine equations.

## V.1. Continued fractions

Here we consider the one variable case which we approach through the classical theory of continued fractions. We just skim through the theory leaving you to prove the main results (those who are lazy could perhaps refer to one of the standard text books such as Hardy and Wright [**96**] or Davenport [**43**]).

Let $\alpha$ denote a real number and $\lfloor x \rfloor$ denote the largest integer less than or equal to $x$, sometimes called the *floor* of $x$. Set $a_0 = \lfloor \alpha \rfloor$. Then if $a_0 \neq \alpha$ define $\alpha_1 > 1$ by $\alpha = a_0 + 1/\alpha_1$ and put $a_1 = \lfloor \alpha_1 \rfloor$. Now if $a_1 \neq \alpha_1$ we define $\alpha_2, a_2$ in a similar manner and so on. We hence obtain a sequence of numbers $a_i$ which is finite if and only if the final term $a_n$ is equal to $\alpha_n$. If such a sequence is finite then it is obvious that $\alpha$ is a rational number.

We also define two other sequences $(p_n), (q_n)$ by the formulae

$$p_n = a_n p_{n-1} + p_{n-2} \ , \ q_n = a_n q_{n-1} + q_{n-2}$$

with initial values given by $p_0 = a_0$, $q_0 = 1$, $p_1 = a_0a_1 + 1$ and $q_1 = a_1$. The $a_n$ are called the partial quotients of $\alpha$ while the fractions $p_n/q_n$ are called the convergents to $\alpha$. The complete quotients of $\alpha$ are the numbers $\alpha_i$. There is another interpretation of the convergents as a 'continued fraction':

$$p_n/q_n = a_0 + \cfrac{1}{a_1 + \cfrac{1}{a_2 + \cfrac{\ddots}{\frac{1}{a_n}}}}$$

The convergents satisfy the following identity which is crucial for the study of continued fractions:

LEMMA V.1. *For $n \in \mathbb{N}$ we have*

$$p_nq_{n-1} - q_np_{n-1} = (-1)^{n-1}.$$

PROOF. We set $\Delta_n = p_nq_{n-1} - q_np_{n-1}$ and prove the result by induction on $n$. The result is clearly true for $n = 1$ as

$$\Delta_1 = (a_0a_1 + 1) - a_0a_1 = 1.$$

Hence we shall assume the result has been proved for all $k < n$, and try to prove the result for $n$:

$$\begin{aligned}\Delta_n &= (a_np_{n-1} + p_{n-2})q_{n-1} - (a_nq_{n-1} + q_{n-2})p_{n-1} \\ &= p_{n-2}q_{n-1} - q_{n-2}p_{n-1} \\ &= -\Delta_{n-1} = (-1)^{n-1}.\end{aligned}$$

So the result follows by induction. □

We note the following identity which shall come in useful later:

$$\alpha = \frac{\alpha_{n+1}p_n + p_{n-1}}{\alpha_{n+1}q_n + q_{n-1}}.$$

This allows us to show

THEOREM V.2. *The $p_n$ and $q_n$ are relatively prime integers such that $p_n/q_n \to \alpha$ as $n \to \infty$.*

PROOF. That the $p_n$ and $q_n$ are relatively prime follows from Lemma V.1. To see convergence notice that

$$\begin{aligned}\alpha - \frac{p_n}{q_n} &= \frac{\alpha_{n+1}p_n + p_{n-1}}{\alpha_{n+1}q_n + q_{n-1}} - \frac{p_n}{q_n} \\ &= \frac{\Delta_n}{q_n(\alpha_{n+1}q_n + q_{n-1})} \\ &= \frac{\pm 1}{q_n(\alpha_{n+1}q_n + q_{n-1})}.\end{aligned}$$

Now as $\alpha_{n+1} > a_{n+1}$ we have

$$\left|\alpha - \frac{p_n}{q_n}\right| \leq \frac{1}{q_n q_{n+1}}.$$

But $(q_n)$ is a strictly increasing sequence of positive real numbers so

$$\lim_{n\to\infty} \frac{p_n}{q_n} = \alpha,$$

as required. □

We write $p_n/q_n = [a_0, a_1, \dots , a_n]$ and

$$\alpha = \lim_{n\to\infty} p_n/q_n = [a_0, a_1, a_2, \dots].$$

As an example the first four convergents to $\pi$ are given by

$$\begin{array}{rcll} [3,7] & = & 22/7, & |7\pi - 22| \leq 10^{-2} \\ [3,7,15] & = & 333/106, & |106\pi - 333| \leq 10^{-2} \\ [3,7,15,1] & = & 355/113, & |113\pi - 355| \leq 10^{-4} \\ [3,7,15,1,292] & = & 103993/33102, & |33102\pi - 103993| \leq 10^{-4} \end{array}$$

The number with the worst behaved convergents (in terms of the rate of convergence) is

$$\alpha = \frac{1}{2}(1+\sqrt{5}) = [1,1,1,1,\dots].$$

We have the following result, the proof of which is left as an exercise:

THEOREM V.3. *The following facts hold for the convergents of an irrational real number, $\alpha$:*

- *The convergents $(p_n, q_n)$ make the linear form $x - y\alpha$ very small. Explicitly we have*
$$|q_n\alpha - p_n| \leq 1/q_n.$$
*Indeed if $p, q$ are two integers such that*
$$|q\alpha - p| \leq \frac{1}{2q}$$
*then $p/q$ is a convergent to $\alpha$.*
- *The denominators, $q_n$, of the convergents, $p_n/q_n$, increase at an exponential rate.*

Note that we have phrased the above in terms of how to make a two term linear form become very small. The generalization of the above to linear forms in many variables will be the main subject of this chapter. We end our discussion of continued fractions by looking at the continued fraction expansion of quadratic irrationals.

A purely periodic continued fraction is a continued fraction for which there exists an integer $n$, called the period, such that

$$a_i = a_{n+1+i} \quad \text{for all } i \geq 0.$$

If $\alpha$ is the limit of such a continued fraction we write

$$\alpha = [\overline{a_0, a_1, \ldots, a_{n-1}, a_n}].$$

In such a situation $\alpha$ satisfies the equation

$$\alpha = \alpha_{n+1} = \frac{\alpha p_n + p_{n-1}}{\alpha q_n + q_{n-1}}.$$

On clearing denominators this means that $\alpha$ must satisfy a quadratic equation. On the other hand we call a real quadratic irrational, $\alpha$, reduced if $\alpha > 1$ and $\alpha'$, the quadratic conjugate of $\alpha$, satisfies $-1 < \alpha' < 0$. Given this definition we can then link purely periodic continued fractions and reduced real quadratic irrationals by

THEOREM V.4. *The continued fraction expansion of a real number $\theta$ is purely periodic if and only if $\theta$ is a reduced real quadratic irrational.*

PROOF. See [**43**][Chapter IV] or one of the exercises at the end of this chapter. □

Now let $N$ denote a positive integer which is not a perfect square. Clearly $\sqrt{N}$ is not a reduced quadratic irrational as $-\sqrt{N} < -1$. However, if we set $a_0 = \lfloor \sqrt{N} \rfloor$ then $a_0 + \sqrt{N}$ is a reduced quadratic irrational. By the above theorem it will have a purely periodic continued fraction expansion, which will be of the form

$$a_0 + \sqrt{N} = [\overline{2a_0, a_1, \ldots, a_n}].$$

Hence the continued fraction expansion of $\sqrt{N}$ will be periodic (but not purely periodic) as it is of the form

$$\sqrt{N} = [a_0, \overline{a_1, a_2, \ldots, a_n, 2a_0}]$$

with an obvious notation. For example we obtain

$$\sqrt{46} = [6, \overline{1, 3, 1, 1, 2, 6, 2, 1, 1, 3, 1, 12}].$$

We can then deduce how to solve Pell's equation

$$x^2 - Ny^2 = 1$$

or the more general equation

$$x^2 - Ny^2 = \pm 1.$$

THEOREM V.5. *Suppose*

$$\sqrt{N} = [a_0, \overline{a_1, a_2, \ldots, a_n, 2a_0}]$$

*and let $k$ denote an integer of the form $m(n+1) - 1$. Then the convergents $(p_k, q_k)$ are solutions to*

$$p_k^2 - Nq_k^2 = \pm 1$$

PROOF. We have

$$\sqrt{N} = \frac{\alpha_{k+1} p_k + p_{k-1}}{\alpha_{k+1} q_k + q_{k-1}}$$

and as $\sqrt{N}$ is periodic we have, by choice of $k$, $\alpha_{k+1} = \sqrt{N} + a_0$. So clearing denominators in the identity above we find

$$(\sqrt{N} + a_0)p_k + p_{k-1} = \sqrt{N}((\sqrt{N} + a_0)q_k + q_{k-1}).$$

Equating coefficients of $\sqrt{N}$ gives us

$$\begin{aligned} p_{k-1} &= Nq_k - a_0 p_k, \\ q_{k-1} &= p_k - a_0 p_k. \end{aligned}$$

But then we have

$$\begin{aligned} (-1)^{k-1} &= p_k q_{k-1} - q_k p_{k-1} \\ &= p_k(p_k - a_0 p_k) - q_k(Nq_k - a_0 p_k) \\ &= p_k^2 - Nq_k^2. \end{aligned}$$

So $(p_k, q_k)$ is a solution to $p_k^2 - Nq_k^2 = \pm 1$. □

It can be shown that the solution to

$$x^2 - Ny^2 = \pm 1$$

given by $(p_n, q_n)$ is the smallest such non-trivial solution and hence

$$p_n + \sqrt{N} q_n$$

is the fundamental unit of the field $\mathbb{Q}(\sqrt{N})$ when $N$ is square free and not congruent to 1 modulo 4. A similar method can be used to determine the fundamental unit when $N \equiv 1 \pmod 4$, see [**32**] for details.

As an example of solving Pell's equation consider

$$x^2 - 13y^2 = \pm 1$$

in this case we have

$$\sqrt{13} = [3, \overline{1, 1, 1, 1, 6}]$$

and with $k = 4$ we find $(p_k, q_k) = (18, 5)$ and

$$18^2 - 13 \cdot 5^2 = -1.$$

But looking at the next period we have

$$\sqrt{13} = [3, \overline{1, 1, 1, 1, 6, 1, 1, 1, 1, 6}]$$

and with $k = 9$ we find $(p_k, q_k) = (649, 180)$ and

$$649^2 - 13 \cdot 180^2 = 1.$$

## V.2. Approximation lattices

We have just seen how we can produce very good rational approximations to real numbers which have 'small' numerators and denominators. We indicated that one could prove that to produce very good approximations you needed to use larger and larger numbers in the numerators and denominators. In this section we shall consider the following generalization.

Consider the linear form

$$L(\vec{x}) = \alpha_0 + x_1\alpha_1 + \cdots + x_n\alpha_n$$

where $\alpha_i \in \mathbb{R}$ are given and the $x_i$ are integer variables. Such a linear form is called homogeneous if $\alpha_0 = 0$, otherwise it is called inhomogeneous. One could ask how small could

$$|L(\vec{x})|$$

become if we restricted the $x_i$ to be not too 'large'. We obviously exclude some rather trivial cases, for example

- In the homogeneous case we do not consider the case $x_i = 0$ for all $i$.
- We assume the $\alpha_i$ are linearly independent over $\mathbb{Z}$. In other words we cannot have $L(\vec{x}) = 0$ with $\vec{x} \in \mathbb{Z}^n \setminus \{\vec{0}\}$.

To study such a situation we encode the linear form $L(\vec{x})$ in a matrix:

$$A = \begin{pmatrix} 1 & & & 0 \\ & \ddots & & \\ 0 & & 1 & 0 \\ \alpha_1 & \dots & \alpha_{n-1} & \alpha_n \end{pmatrix} \in M_{n\times n}(\mathbb{R}).$$

We then look at all vectors of the form $A\vec{x}$, where $\vec{x}$ ranges over all of $\mathbb{Z}^n$. We aim to determine when such vectors are close to the vector

$$(0, \dots, 0, -\alpha_0)^t \in \mathbb{R}^n.$$

The set $\{A\vec{x} : \vec{x} \in \mathbb{Z}^n\}$ is called an approximation lattice in $n$-dimensional euclidean space. In the homogeneous case we are therefore looking for small vectors in the lattice. We have translated our diophantine approximation problem into a problem about lattices. The advantage is that the computational theory of lattices is very well developed.

Notice in the above it is more important that $x_1, \dots, x_{n-1}$ are small than $x_n$. In applications this will not be a problem. However, the $n$th element in all vectors in the lattice will be very, very small in comparison to the $x_i$ for any vector $\vec{x}$ which corresponds to a small value of the linear form. Therefore it is common practice to scale the final row by a suitable weight factor. We choose a 'large' integer $C$ and instead consider the matrix

$$A = \begin{pmatrix} 1 & & & 0 \\ & \ddots & & \\ 0 & & 1 & 0 \\ [C\alpha_1] & \dots & [C\alpha_{n-1}] & [C\alpha_n] \end{pmatrix} \in M_{n\times n}(\mathbb{R}) \qquad \text{(V.1)}$$

where $[r]$ denotes the nearest integer to $r$ (with any fixed convention for numbers of the form $(2m+1)/2$).

The above situation can be generalized to the situation where we have many linear forms. Suppose we have the following set of linear forms:

$$L_j(\vec{x}) = \alpha_{j,0} + x_1\alpha_{j,1} + \cdots + x_n\alpha_{j,n} \ , \ j = 1, \dots, m \leq n.$$

We can then approximate these with the lattice generated by the columns of the following matrix:

$$A = \begin{pmatrix} 1 & & & & 0 \\ & \ddots & & & \\ 0 & & 1 & & \\ [C\alpha_{1,1}] & \dots & [C\alpha_{1,n-m}] & \dots & [C\alpha_{1,n}] \\ \vdots & & \vdots & & \vdots \\ [C\alpha_{m,1}] & \dots & [C\alpha_{m,n-m}] & \dots & [C\alpha_{m,n}] \end{pmatrix} \in M_{n\times n}(\mathbb{R}).$$

Using this we can produce approximation lattices of complex linear forms as follows: Suppose now that the $\alpha_i$ are complex numbers. If $L(\vec{x})$ is small then both the real and imaginary parts of $L(\vec{x})$ should be simultaneously small. Hence we approximate the linear forms $\Re(L(\vec{x})$ and $\Im(L(\vec{x}))$ by the lattice

$$A = \begin{pmatrix} 1 & & & 0 \\ & \ddots & & \\ 0 & & 1 & \\ [C\Re(\alpha_1)] & \dots & [C\Re(\alpha_{n-1})] & [C\Re(\alpha_n)] \\ [C\Im(\alpha_1)] & \dots & [C\Im(\alpha_{n-1})] & [C\Im(\alpha_n)] \end{pmatrix} \in M_{n\times n}(\mathbb{R}).$$

We shall always permute the $\alpha_i$ such that the determinant of the above matrix is non-zero, if possible. We shall not consider cases where this cannot be done.

## V.3. Lattices

We now discuss the basic theory of lattices. Fix a positive integer $n$. A lattice is a $\mathbb{Z}$-module spanned by $n$ linearly independent vectors in $\mathbb{R}^n$. The spanning set is called the basis of the lattice. Hence a lattice is a set

$$\mathcal{L} = \left\{ \sum_{i=1}^{n} x_i\vec{b}_i : x_i \in \mathbb{Z} \right\}$$

with the column vectors $\vec{b}_1, \dots, \vec{b}_n$ being a basis of $\mathcal{L}$ (and hence a basis of $\mathbb{R}^n$). We let $B$ denote the matrix with columns $\vec{b}_1, \dots, \vec{b}_n$. It is easily seen that a lattice basis is unique up to multiplication on the right by an element of $GL_n(\mathbb{Z})$. The determinant of the lattice is defined to be $\Delta(\mathcal{L}) = |\det(B)|$, so the determinant does not depend on the choice of basis.

If we let $\langle\ ,\ \rangle$ denote the standard euclidean inner-product on $\mathbb{R}^n$ then the square of the euclidean norm of an arbitrary element $v = \sum x_i \vec{b}_i$ in the lattice

$$\|\vec{v}\|^2 = \vec{v}^t\vec{v} = \vec{x}^t B^t B \vec{x} = Q(\vec{x})$$

gives a positive definite quadratic form $Q(\vec{x})$.

Similarly every positive definite quadratic form, $Q(\vec{x})$, gives rise to a lattice. If we let $A$ denote the matrix of the quadratic form. Then after computing its Cholesky decomposition $C^tC$ we can take the columns of $C$ as a basis of a lattice.

As the lattice is discrete there is a well defined value for the size of the smallest non-zero vector in the lattice. This value is called the first successive minimum of the lattice, and it is usually denoted $M_1$. Determining the exact value of $M_1$ is a very hard problem. There is an algorithm, due to Fincke and Pohst [**56**] for determining the smallest vectors in a lattice but it is rather slow in practice for large lattices. Later in this chapter we will see that we can find a very good guess for the smallest vector in a lattice using the LLL–algorithm. For now, however, we will content ourselves with the following upper bound on $M_1$.

THEOREM V.6 (Hermite's theorem). *Let $\mathcal{L}$ denote an $n$-dimensional lattice with first successive minimum $M_1$. There exists a constant $\mu_n \in \mathbb{R}^{>0}$ depending only on $n$ such that*

$$M_1^n \le \mu_n \Delta(\mathcal{L})^2.$$

PROOF. See [**155**, pp 197–198]. □

The best possible values of the constants $\mu_n$ of the theorem are called 'Hermite's constants', and are denoted $\gamma_n^n$. Their exact values are only known for $n \le 8$,

$$\begin{array}{llll} \gamma_1^1 = 1, & \gamma_2^2 = \frac{4}{3}, & \gamma_3^3 = 2, & \gamma_4^4 = 4, \\ \gamma_5^5 = 8, & \gamma_6^6 = \frac{64}{3}, & \gamma_7^7 = 64, & \gamma_8^8 = 256, \end{array}$$

while for $n \ge 9$ there is only the upper bound

$$\gamma_n^n \le \left(\frac{2}{\pi}\right)^n \Gamma\left(\frac{n+4}{2}\right)^2.$$

We can consider lattices as analogous to finite dimensional inner product spaces that one meets in linear algebra courses, except the scalars are restricted to lie in the ring $\mathbb{Z}$. The first result one usually learns about inner product spaces is that any basis can be altered to form an orthogonal basis with respect to the inner-product. This is done by means of the Gram–Schmidt process:

THEOREM V.7 (The Gram–Schmidt Process). *A vector space with a basis $\vec{b}_1$, ..., $\vec{b}_n$ and inner product $\langle\ ,\ \rangle$ has an orthogonal basis given by $\vec{b}_1^*$, ..., $\vec{b}_n^*$ where*

$$\vec{b}_i^* = \vec{b}_i - \sum_{j=1}^{i-1} \mu_{i,j}\vec{b}_j^*\ ,\ i = 1, \ldots, n$$

*and the $\mu_{i,j}$ are defined by*

$$\mu_{i,j} = \frac{\langle \vec{b}_i, \vec{b}_j^* \rangle}{\langle \vec{b}_j^*, \vec{b}_j^* \rangle}.$$

In our situation this is useless as we can only make such a change of variable if $\mu_{i,j} \in \mathbb{Z}$ for all $i$ and $j$. The idea behind the algorithm of Lenstra, Lenstra and Lovász is to try and get close to such an orthogonal basis. Orthogonal bases provide one of the main tools in numerical analysis for approximating functions. We shall see that the almost orthogonal bases of Lenstra, Lenstra and Lovász provide a powerful tool when it comes to diophantine approximations.

Following Lenstra, Lenstra and Lovász we call a basis, $B$, of a lattice LLL–reduced if the associated Gram–Schmidt basis $B^*$ and constants $\mu_{i,j}$ satisfy

1.
$$|\mu_{i,j}| \leq 1/2\ ,\ 1 \leq j < i \leq n, \text{ and}$$

2.
$$\|\vec{b}_i^* + \mu_{i,i-1}\vec{b}_{i-1}^*\|^2 \geq \frac{3}{4}\|\vec{b}_{i-1}^*\|^2\ ,\ 1 < i \leq n.$$

The first condition says that the vectors of an LLL–reduced basis are almost orthogonal, the second condition imposes a restriction on the relative sizes of the vectors in the basis, since the second condition can be rewritten as

$$\|\vec{b}_i^*\|^2 \geq \left(\frac{3}{4} - \mu_{i,i-1}^2\right)\|\vec{b}_{i-1}^*\|^2,$$

as can at once be verified using the fact that the Gram–Schmidt vectors, $\vec{b}_i^*$, are orthogonal.

Such a basis always exists and can be computed very quickly, a result which we shall leave to the next section. In the rest of this section we detail some of the properties that an LLL–reduced basis will have, the main result being:

THEOREM V.8. *Assume that the columns of the matrix $B$ represents an LLL–reduced basis of a lattice $\mathcal{L}$. Let the columns of $B^*$ denote the Gram–Schmidt basis constructed from $B$ using the formula above. Then*

1. *For $1 \leq j \leq i \leq n$ we have*

$$\|\vec{b}_j\|^2 \leq 2^{i-1}\|\vec{b}_i^*\|^2.$$

2.

$$\Delta(\mathcal{L}) \leq \prod_{i=1}^{n} \|\vec{b}_i\| \leq 2^{n(n-1)/4}\Delta(\mathcal{L}).$$

3.

$$\|\vec{b}_1\| \leq 2^{(n-1)/4}\Delta(\mathcal{L})^{1/n}$$

Before we prove this result we note that it is really only the last of these properties which we shall use. In applications we shall use it to estimate a good value for the mysterious constant $C$ mentioned earlier, (V.1).

PROOF. 1. Given that the basis is LLL–reduced we find that for $i = 2, \dots, n$,

$$\|\vec{b}_i^*\|^2 \geq \left(\frac{3}{4} - \mu_{i,i-1}^2\right)\|\vec{b}_{i-1}^*\|^2 \geq \frac{1}{2}\|\vec{b}_{i-1}^*\|^2.$$

Hence by induction, for $1 \leq j \leq i \leq n$, the inequality $\|\vec{b}_j^*\|^2 \leq 2^{i-j}\|\vec{b}_i^*\|^2$ holds. Then from the definition of the Gram–Schmidt basis we find

$$\begin{aligned}
\|\vec{b}_i\|^2 &= \|\vec{b}_i^*\|^2 + \sum_{j=1}^{i-1} \mu_{i,j}^2 \|\vec{b}_j^*\|^2, \\
&\leq \left(1 + \sum_{j=1}^{i-1} 2^{i-j-2}\right)\|\vec{b}_i^*\|^2, \\
&= \left(1 + \frac{1}{4}(2^i - 2)\right)\|\vec{b}_i^*\|^2, \\
&\leq 2^{i-1}\|\vec{b}_i^*\|^2.
\end{aligned}$$

So we obtain for $1 \leq j \leq i \leq n$,

$$\|\vec{b}_j\|^2 \leq 2^{j-1}\|\vec{b}_j^*\|^2 \leq 2^{j-1+i-j}\|\vec{b}_i^*\|^2 \leq 2^{i-1}\|\vec{b}_i^*\|^2.$$

Which is the first statement we were required to prove.

2. As the $\vec{b}_i^*$ are orthogonal it is clear that we have

$$\Delta(\mathcal{L}) = |\det(\vec{b}_1^*, \dots, \vec{b}_n^*)| = \prod_{i=1}^{n} \|\vec{b}_i^*\|.$$

From the first statement of the theorem, which we have just proved, and the fact that $||\vec{b}_i^*|| \leq ||\vec{b}_i||$ we obtain

$$\begin{aligned} \Delta(\mathcal{L}) &\leq \prod_{i=1}^{n} ||\vec{b}_i||, \\ &\leq \prod_{i=1}^{n} 2^{(i-1)/2} ||\vec{b}_i^*||, \\ &\leq 2^{n(n-1)/4} \prod_{i=1}^{n} ||\vec{b}_i^*||, \\ &= 2^{n(n-1)/4} \Delta(\mathcal{L}). \end{aligned}$$

Which proves the second statement of the theorem.

3. To prove the third and final statement set $j = 1$ into the first statement and take the product over all possible $i$ to obtain

$$\begin{aligned} ||\vec{b}_1||^{2n} &\leq \prod_{i=1}^{n} 2^{i-1} ||\vec{b}_i^*||^2, \\ &\leq 2^{n(n-1)/2} \Delta(\mathcal{L})^2. \end{aligned}$$

Which concludes the proof of our three results. □

For our applications we shall use the fact that the first element in an LLL–reduced basis provides a very good guess for the size of the smallest vector in the lattice. This is quantified by the next result:

THEOREM V.9. *Let $B$ be a reduced basis for a lattice $\mathcal{L}$, then for all $\vec{x} \neq \vec{0}$ in the lattice $\mathcal{L}$ we have*

$$||\vec{b}_1||^2 \leq c_1 ||\vec{x}||^2,$$

*with $c_1$ given by*

$$c_1 = \max\{||\vec{b}_1||^2 / ||\vec{b}_i^*||^2 : 1 \leq i \leq n\}.$$

PROOF. We certainly have that $||\vec{b}_1||^2 \leq c_1 ||\vec{b}_i^*||^2$, for all values of $i$. Now write

$$\vec{x} = \sum_{i=1}^{n} r_i \vec{b}_i = \sum_{i=1}^{n} r_i' \vec{b}_i^*,$$

where $r_i \in \mathbb{Z}$ and $r_i' \in \mathbb{R}$. We let $i_0$ denote the largest index with $r_i \neq 0$ then $r_i' = r_i$ and so

$$\begin{aligned} ||\vec{x}||^2 &= \sum_{i=1}^{n} r_i'^2 ||\vec{b}_i^*||^2 \geq r_{i_0}'^{\,2} ||\vec{b}_{i_0}^*||^2 \geq ||\vec{b}_{i_0}^*||^2 \\ &\geq c_1^{-1} ||\vec{b}_1||^2. \end{aligned}$$

And the result follows. □

This result is usually proved with $c_1 = 2^{n-1}$. However, as Lenstra, Lenstra and Lovász point out [**117**] the above is also true. The reason for using the above version is that although $c_1 \leq 2^{n-1}$, in practice this upper bound is much, much too pessimistic. If one actually has an LLL–reduced basis then computing the value of $c_1$ above is trivial. In many cases it actually comes out to be very close to 1.0 and hence $\vec{b}_1$ is the smallest vector in the lattice, or very close to it.

The final result of this section estimates the distance from a vector which is not in the lattice to a lattice vector. This will be of use later when we use lattices to approximate inhomogeneous linear forms. We first suppose we have a vector $\vec{y}$ such that $\vec{y} \notin \mathcal{L}$. We then define the vector $\vec{\sigma} \in \mathbb{R}^n$ by $\vec{\sigma} = B^{-1}\vec{y}$, where $B$ denotes the basis matrix of an LLL–reduced basis for the lattice $\mathcal{L}$. In what follows, the symbol $\{x\}$, for a real number $x$, will denote the distance from $x$ to the nearest integer.

THEOREM V.10 ([**208**]). *Let $i_0$ denote the largest index such that $\{\sigma_{i_0}\} \neq 0$. Then for all $\vec{x} \in \mathcal{L}$ we have*

$$||\vec{x} - \vec{y}||^2 \geq c_1^{-1}\{\sigma_{i_0}\}||\vec{b}_1||^2.$$

PROOF. As above we write

$$\vec{x} = \sum_{i=1}^{n} r_i\vec{b}_i = \sum_{i=1}^{n} r_i'\vec{b}_i^*.$$

Similarly we define $\sigma_i'$ by

$$\vec{y} = \sum_{i=1}^{n} \sigma_i\vec{b}_i = \sum_{i=1}^{n} \sigma_i'\vec{b}_i^*.$$

Let $i_1$ be the largest integer such that $r_{i_1} \neq \sigma_{i_1}$; then $r_{i_1} - \sigma_{i_1} = r_{i_1}' - \sigma_{i_1}'$. We then have that

$$||\vec{x} - \vec{y}||^2 \geq |r_{i_1} - \sigma_{i_1}|^2||\vec{b}_{i_1}^*||^2 \geq |r_{i_1} - \sigma_{i_1}|^2 c_1^{-1}||\vec{b}_1||^2.$$

Now if $i_1 < i_0$ then $\sigma_{i_0} = r_{i_0} \in \mathbb{Z}$, which is a contradiction. If $i_1 = i_0$ then we have $|r_{i_1} - \sigma_{i_1}| = |r_{i_0} - \sigma_{i_0}| \geq \{\sigma_{i_0}\}$, and we are done. Finally if $i_1 > i_0$ then $\sigma_{i_1} \in \mathbb{Z}$ and as $\sigma_{i_1} \neq r_{i_1}$ and $\{\sigma_{i_0}\} \leq 1/2$ we have that $|r_{i_1} - \sigma_{i_1}| \geq 1 \geq \{\sigma_{i_0}\}$. □

So the previous two theorems give us a way of computing a lower bound on the following quantity

$$\ell(\mathcal{L}, \vec{y}) = \begin{cases} \min\limits_{\vec{x} \in \mathcal{L}} ||\vec{x} - \vec{y}|| & \vec{y} \notin \mathcal{L} \\ \min\limits_{\vec{0} \neq \vec{x} \in \mathcal{L}} ||\vec{x}|| & \vec{y} \in \mathcal{L} \end{cases}$$

## V.4. The LLL–algorithm

Below we present the algorithm. First we require two subprocedures: In what follows the columns of the matrix $B$ represent our basis vectors, the elements of the matrix $U$ represent the entries $\mu_{i,j}$ from the Gram–Schmidt process and the vector $\mathfrak{B}$ contains the inner products $\langle \vec{b}_i^*, \vec{b}_i^* \rangle$.

---

**Procedure A**

---

DESCRIPTION: Make the entry $u_{k,l}$ satisfy $|u_{k,l}| \leq \frac{1}{2}$.
INPUT: Integers $k, l$; Square $n \times n$ Matrices $B, U$.
OUTPUT: Same.

If $(|\mu_{k,l}| > 1/2)$ then
$r := [\mu_{k,l}]$.
$\vec{b}_k := \vec{b}_k - r\vec{b}_l$.
For $j = 1$ to $l - 1$ do $\mu_{k,j} := \mu_{k,j} - r\mu_{l,j}$.
$\mu_{k,l} := \mu_{k,l} - r$
Endif.

---

Note on exit from this procedure we have that $\mu_{k,l} \leq 1/2$, as required.

---

**Procedure B**

---

DESCRIPTION: Interchange $\vec{b}_k$ and $\vec{b}_{k-1}$.
INPUT: Integer $k$; Square $n \times n$ Matrices $B, U$; Vector $\mathfrak{B}$.
OUTPUT: Same.

Set $u := \mu_{k,k-1}$; $\mathcal{B} := \mathfrak{B}_k + u^2\mathfrak{B}_{k-1}$; $\mu_{k,k-1} := u\mathfrak{B}_{k-1}/\mathcal{B}$;
$\mathfrak{B}_k := \mathfrak{B}_{k-1}\mathfrak{B}_k/\mathcal{B}$; $\mathfrak{B}_{k-1} := \mathcal{B}$;
Swap the vectors $\vec{b}_{k-1}$ and $\vec{b}_k$.
For $j = 1$ to $k - 2$ swap the elements $\mu_{k-1,j}$ and $\mu_{k,j}$.
For $i = k + 1$ to $n$ do
$t := \mu_{i,k-1}$,
$\mu_{i,k-1} := \mu_{k,k-1}\mu_{i,k-1} + \mu_{i,k} - \mu_{i,k}\mu_{k,k-1}u$.
$\mu_{i,k} := t - u\mu_{i,k}$.
Enddo.

---

If we define, for $i = 1, \ldots, n$,

$$D_i = \det(\langle \vec{b}_j, \vec{b}_l \rangle_{1 \leq j,l \leq i}) = \prod_{j=1}^{i} \langle \vec{b}_j^*, \vec{b}_j^* \rangle$$

then if on entering Procedure B we have $\mathfrak{B}_k < (\frac{3}{4} - \mu_{k,k-1}^2)\mathfrak{B}_{k-1}$ then the value of $D_{k-1}$ is reduced by a factor of less than $3/4$. The other $D_i$, however, remain unchanged. Now we can present the main LLL–reduction algorithm.

**LLL–algorithm**

DESCRIPTION: On input of any given basis $B$ the procedure finds a reduced basis.

INPUT: Matrix $B$ whose columns represent the basis of the lattice, $\mathcal{L}$.

OUTPUT: Matrix $B$ whose columns represent the reduced basis of the lattice, $\mathcal{L}$.

```
Compute the $\mu_{i,j}$ and $\vec{b}_i^*$ from the Gram--Schmidt process.
Set for $i = 1$ to $n$, set $\mathfrak{B}_i := \langle \vec{b}_i^*, \vec{b}_i^* \rangle$.
$k := 2$.
Repeat the following:
    Perform Procedure A for $l = k - 1$.
    If $(\mathfrak{B}_k < (\frac{3}{4} - \mu_{k,k-1}^2)\mathfrak{B}_{k-1})$ then
        Perform Procedure B
        If $k > 2$ then $k := k - 1$.
    Else
        For $l = k - 2$ to 1 perform Procedure A.
        $k := k + 1$.
    Endif.
Until $(k > n)$.
```

It is clear that the above algorithm will give us a reduced lattice if we can show that it terminates. To show termination set

$$D = \prod_{i=1}^{n-1} D_i$$

then $D$ is only changed by the algorithm on passing through Procedure B where it is decreased by a value of less than $3/4$. If there is a positive lower bound for $D$ depending only on the lattice then Procedure B can only be called a finite number of times and the program will terminate. That $D$ has a positive lower bound will follow from:

LEMMA V.11. *There is a positive constant $M_1$ depending only on $\mathcal{L}$ such that*

$$D_i \geq \left(\frac{M_1}{\gamma_i}\right)^i.$$

PROOF. Let $\mathcal{L}_i$ denote the sublattice spanned by $\vec{b}_1, \ldots, \vec{b}_i$, then $D_i$ denotes the square of the determinant of the lattice $\mathcal{L}_i$. Now by Hermite's theorem, $\mathcal{L}_i$ contains a non-zero lattice vector with $\|\vec{x}\| \leq \gamma_i D_i^{1/i}$. But there is a constant, $M_{1,i}$, the first successive minimum of $\mathcal{L}_i$, depending only on $\mathcal{L}$ such that $\|\vec{x}\| \geq M_{1,i}$. The result then follows as $M_{1,i} \geq M_1$ the first successive minimum of $\mathcal{L}$. □

As can be seen from our discussion of approximation lattices above we shall mainly be interested in using the LLL–algorithm on lattices which lie in $\mathbb{Z}^n$. Using the algorithm in its original form above could lead to some trouble. If we work with real approximations then it is not clear what accuracy we need to take to guarantee our results are correct. If we work with exact rational arithmetic the numerators and denominators we encounter may 'blow up', i.e. they may become prohibitively large. It is to overcome these problems that we now present de Weger's variant [**207**] of the LLL–algorithm, which only uses integer arithmetic. It can be shown that the following algorithm runs in polynomial time and that the numbers involved do not suffer from coefficient swell.

We firstly have to compute the Gram–Schmidt vectors without using any divisions, which may lead to non-integral results. We assume that the initial matrix $B$ has only integral entries. As above, set $D_i = \det(\langle \vec{b}_j, \vec{b}_l \rangle_{1 \leq j,l \leq i})$; we shall use these $D_i$ as the denominators for our algorithm as

$$\vec{c}_i = D_{i-1}\vec{b}_i^* \in \mathbb{Z}^n \ , \ \lambda_{i,j} = D_j \mu_{i,j} \in \mathbb{Z},$$

where the $\mu_{i,j}$ are the coefficients from the Gram–Schmidt process. We are then led to the following initialization step for an integral version of the LLL–algorithm.

---

**Procedure INIT**

---

```
DESCRIPTION: Computes the Gram--Schmidt basis using no rational
             arithmetic.
INPUT:       Square n × n Matrix B.
OUTPUT:      Square n × n Matrix Λ; Vector D.
   D_0 := 1.
   For i = 1 to n do
       c_i := b_i.
       For j = 1 to i − 1 do
           λ_{i,j} := ⟨b_i, c_j⟩.
           c_i := (D_j c_i − λ_{i,j} c_j)/D_{j−1}.
       Enddo.
       D_i := ⟨c_i, c_i⟩/D_{i−1}.
   Enddo.
```

---

Procedures A and B are much as before except we have to carry around the vector $D$ holding the denominators.

**Procedure A′**

DESCRIPTION: Make the entry $u_{k,l}$ satisfy $|u_{k,l}| \le \frac{1}{2}$.
INPUT: Integers $k, l$; Square $n \times n$ Matrices $B, \Lambda$; Vector $D$.
OUTPUT: Same.

If $(2|\lambda_{k,l}| > D_l)$ then
  $r := [\lambda_{k,l}/D_l]$.
  $\vec{b}_k := \vec{b}_k - r\vec{b}_l$.
  For $j = 1$ to $l - 1$ do $\lambda_{k,j} := \lambda_{k,j} - r\lambda_{l,j}$.
  $\lambda_{k,l} := \lambda_{k,l} - rD_l$
endif.

**Procedure B′**

DESCRIPTION: Interchange $\vec{b}_k$ and $\vec{b}_{k-1}$.
INPUT: Integer $k$; Square $n \times n$ Matrices $B, \Lambda$; Vector $D$.
OUTPUT: Same.

Swap the vectors $\vec{b}_{k-1}$ and $\vec{b}_k$.
For $j = 1$ to $k - 2$ swap the elements $\lambda_{k-1,j}$ and $\lambda_{k,j}$.
For $i = k + 1$ to $n$ do
  $t := \lambda_{i,k-1}$,
  $\lambda_{i,k-1} := (\lambda_{i,k-1}\lambda_{k,k-1} + \lambda_{i,k}D_{k-2})/D_{k-1}$.
  $\lambda_{i,k} := (tD_k - \lambda_{i,k}\lambda_{k,k-1})/D_{k-1}$.
Enddo.
$D_{k-1} := (D_{k-2}D_k + \lambda_{k,k-1}^2)/D_{k-1}$.

We can now present De Weger's algorithm. It is left as an exercise to check that no rational numbers will occur during its execution. You only need to check the procedures INIT and B′.

**De Weger's LLL–algorithm**

DESCRIPTION: On input of a set of integral basis elements the procedure computes an integral LLL--reduced basis.
INPUT: Integral matrix $B$ whose columns represent the basis of the lattice, $\mathcal{L}$.
OUTPUT: Integral matrix $B$ whose columns represent the reduced basis of the lattice, $\mathcal{L}$.

Perform Procedure INIT.
$k := 2$.

```
Repeat the following:
     Perform Procedure A' for $l = k - 1$.
     If $4D_{k-2}D_k < (3D_{k-1}^2 - 4\lambda_{k,k-1}^2)$ then
          Perform Procedure B'.
          If $k > 2$ then $k := k - 1$.
     Else
          For $l = k - 2$ to 1 perform Procedure A'.
          $k := k + 1$.
     Endif
 Until $(k > n)$.
```

The LLL–algorithm was originally developed in the context of an application to factor polynomials with integer coefficients in polynomial time [**117**]. This application was extended to factoring polynomials over algebraic number fields [**115**]. However, since then it has become widely used in all areas of computational number theory, see [**32**].

The definition of an LLL–reduced basis can be slightly modified by altering the constant $3/4$ to any number, $\omega$, in the interval $(0.25, 1.0)$; of course the algorithm will then need altering. The larger the value of $\omega$ then the better behaved the basis vectors should be, for example one should get a first basis vector which is much closer to the minimal vector in the lattice. However, the larger the value of $\omega$ then the longer the algorithm will take before it terminates. The choice of $\omega = 3/4$ is standard and appears a good compromise in the situations that we are interested in.

One can also add lines into the LLL–algorithm which keep track of the transformation matrix which produces the reduced basis. This is often all that is required for an application.

There is also a modification of the LLL–algorithm which only has as input the Gram matrix of the basis of the lattice. The Gram matrix, $G$, is the matrix of inner–products of the basis, in other words $G = B^t B$, where $B$ is a matrix whose columns represent the basis of the lattice. In such a situation we only keep track of the transformation matrix and do not worry about the basis matrix at all.

For more advanced versions of the LLL–algorithm, which we will not need in this book, see [**32**], [**164**] and [**51**].

## V.5. Exercises

1). Adapt both variants of the LLL–algorithm so that they also output the transition matrix.

2). Prove Theorem V.3.

3). Show that the constant $c_1$ in Theorem V.9 is bounded above by $2^{n-1}$.

4). Prove the correctness of de Weger's variant of the LLL–algorithm.

5). Show that every positive definite quadratic form can be written as a sum of squares with positive coefficients. Hence give an algorithm to determine the smallest vector in a lattice.

6). Let $\Lambda = x_1\theta_1 + x_2\theta_2$ where $\theta_i \in \mathbb{R}$ and assume that there exist positive constants $c$ and $d$ such that

$$|\Lambda| < ce^{-dX}$$

where $X = \max(|x_1|, |x_2|) < X_0$, where $X_0$ is some given positive real constant. Show how one can deduce a new upper bound on $X$ which is usually much smaller than $X_0$ using continued fractions (and no application of LLL), if $X_0$ is very large.

7). In this exercise we shall prove that a continued fraction is purely periodic if and only if it is the continued fraction of a reduced quadratic irrational.

(i) Let $\alpha$ denote a reduced quadratic irrational. Show that this can be written in the form

$$\frac{P + \sqrt{D}}{Q}$$

where $D$ is positive and not a perfect square and $P$ and $Q$ come from a finite set of positive integers.

(ii) Show that the complete quotients of $\alpha$ are also reduced quadratic irrationals and are of the form

$$\frac{P_1 + \sqrt{D}}{Q_1}.$$

Hence deduce that the continued fraction expansion of $\alpha$ must be eventually periodic.

(iii) Show that for such a continued fraction expansion

$$\alpha_n = \alpha_m$$

implies that

$$\alpha_{n-1} = \alpha_{m-1}.$$

Hence deduce that the continued fraction expansion of a reduced quadratic irrational is purely periodic.

(iv) Show that if a continued fraction expansion of a real number is purely periodic then it is the expansion of a reduced quadratic irrational.

CHAPTER VI

# Applications of the LLL–algorithm

We shall now concentrate on three applications of the LLL–algorithm. The first we give just as a bit of fun. We then turn to show how to use LLL to solve subset-sum problems. Subset-sum (or knapsack) problems are known to belong to the class of NP-complete problems, hence they are considered to be very hard in practice to solve. They are more than just of theoretical interest as one can build public-key cryptosystems from knapsack problems. We shall show that you can often break such a cryptosystem using the LLL–algorithm.

Finally we turn our attention to determining whether a linear form can become exponentially small. It is this last application which forms the backbone of the method to solve diophantine equations via Baker's theory of linear forms in logarithms. The LLL–algorithm reduces the astronomical bounds from Baker's theory to something more manageable.

## VI.1. A 'fun' application

We have seen, in V.1, how to produce rational numbers, $p/q$, with small numerator and denominator which are close to $\pi$. We can think of this as finding polynomials of degree one, i.e. $qX - p$, with small height and with a root close to $\pi$. One natural generalization of this would be to try and look for polynomials of higher degree, with integer coefficients of small height, and which have a root very close to $\pi$. We can do this using the LLL–algorithm and approximation lattices. Obviously we can never find such a polynomial with root exactly equal to $\pi$ as $\pi$ is a transcendental number.

Suppose we wish to look for polynomials of degree 3 with coefficients of the order of $10^2$ and which possess a root very close to $\pi$. One way to look at this is to find a very small value of the linear form

$$|x_1\pi^3 + x_2\pi^2 + x_3\pi + x_4|$$

where we want the above to be small but the $x_i$ to be of order $10^2$. To find such a polynomial we form the approximation lattice generated by the columns of the matrix

$$A = \begin{pmatrix} 1 & 0 & 0 & 0 \\ 0 & 1 & 0 & 0 \\ 0 & 0 & 1 & 0 \\ 3101 & 987 & 314 & 100 \end{pmatrix},$$

where the last row is given by $[100\pi^{4-i}]$. Clearly a small vector in the lattice generated by the columns of $A$ will correspond to a degree 3 polynomial with small coefficients and with a root close to $\pi$. If we find an LLL–reduced basis of this lattice then the first element should correspond to an approximation to the smallest vector in the lattice owing to Theorem V.9. From this we should be able to compute our desired polynomial. We find, using a computer, that an LLL–reduced basis of the lattice spanned by the columns of $A$ is given by $AU$ where

$$U = \begin{pmatrix} -1 & -1 & -2 & -3 \\ 0 & 1 & 2 & -6 \\ 0 & 1 & -5 & 2 \\ 31 & 18 & 58 & 146 \end{pmatrix}.$$

From the first column of $U$ we read off the cubic polynomial $x^3 - 31$ which has a root, $\alpha$, very close to $\pi$. In fact we have $|\alpha - \pi| \leq 0.00022$.

Suppose we need a cubic polynomial with a root closer to $\pi$, then we replace the $A$ above with the following matrix:

$$A = \begin{pmatrix} 1 & 0 & 0 & 0 \\ 0 & 1 & 0 & 0 \\ 0 & 0 & 1 & 0 \\ 31006 & 9870 & 3142 & 1000 \end{pmatrix}.$$

This time we find an LLL–reduced basis is given by $AU$ where

$$U = \begin{pmatrix} 2 & -1 & 3 & 5 \\ -1 & 0 & -1 & 9 \\ -1 & 0 & 6 & 1 \\ -49 & 31 & -102 & -247 \end{pmatrix}.$$

The first column of $U$ then gives us the polynomial $2x^3 - x^2 - x - 49$ which again has a root, $\alpha$, close to $\pi$, indeed $|\alpha - \pi| \leq 0.000027$.

Now suppose we would like a cubic polynomial with one root close to $\pi$ and one root close to $e = 2.71828\ldots$. We would then look at an approximation lattice generated by the columns of a matrix like

$$A = \begin{pmatrix} 1 & 0 & 0 & 0 \\ 0 & 1 & 0 & 0 \\ 3101 & 987 & 314 & 100 \\ 2009 & 739 & 272 & 100 \end{pmatrix}.$$

We then find an LLL–reduced basis is given by $AU$, where the first column of $U$ corresponds to the polynomial $f(x) = 3x^3 - 2x^2 - 66x + 134$. Two of the roots of $f(x)$ are given by approximately 3.14788 and 2.72532. If we wanted a better approximation we could increase the weight given to the last two rows of the matrix $A$ (and hence increase the size of the coefficients of our polynomials), or we could increase the dimension of our matrices and hence the degree of our polynomials.

## VI.2. Knapsack problems

Consider the following problem. We are given a knapsack which can contain certain weights up to a given limit. We are also given a set of objects of various given weights. We are then asked to pack the knapsack so that no more room is left inside it (if possible).

For example suppose the given weights are $1, 2, 4, 8, 16, 32, 64$ and the knapsack can hold up to a weight of 12. We can put weights totaling 12 into the knapsack in exactly one way; namely we put in the weights 4 and 8. Indeed any knapsack total between 1 and 127 can be represented uniquely using such weights and we can determine which weights to use in a quick straightforward manner. This is what is called an *easy* knapsack problem.

Sometimes there may be many solutions to a knapsack problem and sometimes there may be none. If there are no exact solutions then we must search for the best solution possible. So if we have a total knapsack weight of $N$ and $n$ weights each of weight $w_i$ then we need to make

$$\left|\sum_{i=1}^{n} a_i w_i - N\right|$$

as small as possible, where $a_i \in \{0, 1\}$. So all we need to do is solve a linear diophantine equation.

Solving knapsack problems is known to be a very hard problem. It is known to be NP-complete. The fact that it is a very hard problem to solve has led it to be proposed as a scheme for encrypting messages, as we shall now explain.

Suppose we have a set of weights e.g. $w_i = 2^i$, such that for any given number we know in advance that every number in a certain range has at most one representation as a sum of the weights. In addition suppose we wish to transmit a binary message, e.g. $[1, 0, 0, 1, 0]$, we could then send the number

$$1w_0 + 0w_1 + 0w_2 + 1w_3 + 0w_4 \ \left(= 9 \text{ if } w_i = 2^i\right).$$

The receiver has only to know the knapsack weights and then solve the knapsack problem. But solving the knapsack problem for the weights $2^i$ is easy.

To make a public key system we want to publish the weights so that anybody could send us a message. Knowing the weights we should be able to decrypt the message by solving a knapsack problem. However, any hacker could do likewise, hence we need some method which makes it easy for me to solve the knapsack problem but hard for anyone else to.

We do this as follows:

- Choose some easy knapsack weights, say $w_i = 2^i$ for $i = 1, \ldots, K$.
- Then find two coprime integers $N$ and $e$, such that $N$ satisfies

$$N > \sum_{i=1}^{K} w_i.$$

- Compute some hard knapsack weights by the formula
$$h_i \equiv e w_i \pmod{N}$$
and publish the $h_i$ in a table for use as public keys.
- Compute a decrypt key, $d$, such that $de \equiv 1 \pmod{N}$. The value of $d$ can be computed using the extended euclidean algorithm applied to $e$ and $N$.

Now suppose someone wants to send you a message. He, or she, looks up your hard knapsack weights, $h_i$, they then compute their encrypted message;
$$M = \sum_{i=1}^{K} b_i h_i \ , \ b_i \in \{0,1\},$$
where the $b_i$ represent the binary message to be sent. You can decrypt this message as you can compute
$$dM = \sum_{i=1}^{K} b_i(dh_i) \equiv \sum_{i=1}^{K} b_i w_i \pmod{N} = \sum_{i=1}^{K} b_i w_i \left( \text{as } N > \sum_{i=1}^{K} w_i \right).$$
Hence, as the $w_i$ form an easy knapsack problem, we can determine the message, $b_i$.

For example put $w_i = 3^i$, for $i = 1, \ldots, 5$. Now choose
$$N = 400 > 363 = \sum_{i=1}^{5} w_i$$
and let $e = 147$. The decryption key can then be computed to be $d = 283$. Our hard knapsack weights are given by $h_i \equiv 147 \cdot 3^i \pmod{400}$, i.e.
$$\begin{aligned} h_1 &= 41, \quad h_2 = 123, \quad h_3 = 369, \\ h_4 &= 307, \quad h_5 = 121. \end{aligned}$$
Our friend wishes to send us the binary message $[1,0,0,1,1]$. This our friend can encode as
$$41 + 307 + 121 = 469,$$
which they transmit to us. We receive the number 469 and compute
$$283 \times 469 \equiv 327 \pmod{400}.$$
But it is easy for us to solve the easy knapsack problem
$$327 = 1 \cdot 3 + 0 \cdot 3^2 + 0 \cdot 3^3 + 1 \cdot 3^4 + 1 \cdot 3^5.$$
Hence we have recovered the original message.

All that a hacker has to do is to find the original binary sequence given the weights $h_i$ and the encoded message 469. They know neither the original easy knapsack nor the encryption/decryption keys $e, d$ nor even the modulus $N$. Such a problem should be very, very hard to solve; however, it is not as hard as it at first seems as we shall now show.

LLL allows us to break 'low density' knapsack cryptosystems. Basically if we wish to solve the knapsack problem with weights $w_i$ and coefficients $x_i \in \{0, 1\}$ such that

$$\sum_{i=1}^{n} x_i w_i = S$$

then LLL will do it [**143**] if

- $\sum x_i < n/2$,
- $w_i \geq\approx 2^{1.54n}$.

To see how to do this consider the following problem. Using the previous scheme I have created a 'hard' knapsack problem with the weights

$$\begin{array}{lll}
w_1 = 1527086619781 & w_2 = 7635433098905 & w_3 = 14335307584368 \\
w_4 = 150964191369 & w_5 = 754820956845 & w_6 = 3774104784225 \\
w_7 = 18870523921125 & w_8 = 22827045875154 & w_9 = 18767797735142 \\
w_{10} = 22313414945239 & w_{11} = 16199643085567 & w_{12} = 9472641697364 \\
w_{13} = 23521350576663 & w_{14} = 22239321242687 & w_{15} = 15829174572807 \\
w_{16} = 7620299133564 & w_{17} = 14259637757663 & w_{18} = 23614472968001 \\
w_{19} = 22704933199377 & w_{20} = 18157234356257 &
\end{array}$$

I have kept the modulus, the encryption/decryption keys and the original easy knapsack weights secret. Say you wish to send me the binary message

$$1, 0, 0, 0, 1, 1, 0, 0, 0, 1, 1, 0, 0, 0, 1, 1, 0, 0, 0, 1$$

which you encode as the number

$$S = 86175778454285.$$

The question is, can an attacker use LLL to recover your message. Well the answer is yes. He does not know whether the first of the above two conditions are satisfied but he could have a go just in case. The second condition certainly is satisfied so it is certainly worth a try.

Our hacker forms the matrix

$$\begin{pmatrix} 1 & & & & 0 \\ 0 & \ddots & & & \vdots \\ \vdots & & \ddots & & \vdots \\ 0 & \dots & 0 & 1 & 0 \\ w_1 & w_2 & \dots & w_n & -S \end{pmatrix}$$

and computes an LLL–reduced basis of the lattice generated by the columns of this matrix. This is rather easy; it takes a computer less than a second to do this. If you work this out you will see that the first basis vector of the LLL–reduced basis is

$$(1, 0, 0, 0, 1, 1, 0, 0, 0, 1, 1, 0, 0, 0, 1, 1, 0, 0, 0, 1, 0).$$

Hence our hacker has recovered your original message.

## VI.3. Approximating linear forms

We shall mainly be interested, for our later applications, in solving the following problem. Suppose we are given $\alpha_i \in \mathbb{C}$ for $i = 0, \ldots, n$ and two positive real constants $c_2, c_3$. We wish to deduce an upper bound for $H$ in the inequality

$$|\alpha_0 + \sum_{i=1}^{n} x_i \alpha_i| \leq c_2 e^{-c_3 H^q},$$

for some positive integer $q$, given that the integers $x_1, \ldots, x_n$ are bounded by $|x_i| \leq X_i$, where the $X_i$ are given large constants. We would like in such a situation to be able to deduce that $H \leq O(\sqrt[q]{\log X_0})$, where $X_0 = \max X_i$. In other words we would like to show that the linear form cannot become too small if the $|x_i|$ are bounded by the constants $X_i$.

The first non-trivial case is when there are two variables. This was studied in the ground breaking paper of Baker and Davenport [**5**]. This was the first paper to give a strategy to reduce the bounds derived from Baker's theory of linear forms in logarithms. There are a variety of ways to generalize the method of Baker and Davenport. We shall explain the method which has been most successful in recent years.

The basic idea in such a situation, due to de Weger, is to approximate the linear form by an approximation lattice such as those discussed above. We then find a reduced basis for the lattice. The first element in such a basis gives us a close approximation to the shortest vector in the lattice and hence when the linear form could become suitably small. From this we are able to deduce the kind of result we are after. In applications the size of $X_0$ will be of the size of $10^n$; hence deducing a bound on $H$ of the order of $n$ is a very significant result.

We consider three cases

### VI.3.1. Case 1: $\Im(\alpha_i) = 0$ for all $i$.

We choose a constant $C$ of the about the size of $X_0^n$, and then consider the lattice, $\mathcal{L}$, generated by the columns of the matrix

$$\mathcal{A} = \begin{pmatrix} 1 & & & 0 \\ & \ddots & & \\ 0 & & 1 & 0 \\ [C\alpha_1] & \ldots & [C\alpha_{n-1}] & [C\alpha_n] \end{pmatrix} \in \mathbb{Z}^{n \times n}.$$

Note that $C$ has been chosen to make the above lattice have determinant around $X_0^n$ and hence by Theorem V.8 we can hope that the first basis element in an LLL–reduced lattice has order $X_0$. Using de Weger's LLL–algorithm and Theorems V.9 and V.10 we can find, in polynomial time, a lower bound,

$c_4$, on $\ell(\mathcal{L}, \vec{y})$ where

$$\vec{y} = \begin{pmatrix} 0 \\ \vdots \\ 0 \\ -[C\alpha_0] \end{pmatrix} \in \mathbb{Z}^n.$$

If we are lucky we can apply the next lemma to find a bound on $H$ of the order we are looking for.

LEMMA VI.1. *Set $S$ to be the integer $\sum_{i=1}^{n-1} X_i^2$ and $T = (1 + \sum_{i=1}^{n} X_i)/2$. If $c_4^2 \geq T^2 + S$ then either*

$$H \leq \sqrt[q]{\frac{1}{c_3}\left(\log(Cc_2) - \log\left(\sqrt{c_4^2 - S} - T\right)\right)}$$

*or $x_1 = \cdots = x_{n-1} = 0$ and $x_n = -[C\alpha_0]/[C\alpha_n]$.*

Note that the bound on $H$ (if it exists) is of the form

$$O(\sqrt[q]{\log(Cc_2)}) = O(\sqrt[q]{\log(X_0^n c_2)}) = O(\sqrt[q]{\log(X_0)}).$$

PROOF. Put

$$\Phi = [C\alpha_0] + \sum_{i=1}^{n} x_i\,[C\alpha_i]$$

then notice that

$$|\Phi - C(\alpha_0 + \sum_{i=1}^{n} x_i\alpha_i)| \leq \sum_{i=1}^{n} X_i/2 + 1/2 = T,$$

which implies that

$$|\Phi| \leq T + Cc_2 e^{-c_3 H^q}.$$

Now consider the lattice point $\vec{x} = \mathcal{A}\vec{z}$ where

$$\vec{z} = \begin{pmatrix} x_1 \\ \vdots \\ x_n \end{pmatrix};$$

then

$$\vec{x} - \vec{y} = \begin{pmatrix} x_1 \\ \vdots \\ x_{n-1} \\ \Phi \end{pmatrix}.$$

Therefore, either $\vec{x} = \vec{y}$ or

$$\begin{aligned} c_4^2 &\leq \ell(\mathcal{L}, \vec{y})^2 \leq \sum_{i=1}^{n-1} x_i^2 + \Phi^2, \\ &\leq S + \left(T + Cc_2 e^{-c_3 H^q}\right)^2. \end{aligned}$$

Now by assumption $c_4^2 \geq S$ so we have

$$e^{-c_3 H^q} \geq \frac{1}{Cc_2}\left(\sqrt{c_4^2 - S} - T\right).$$

Now again by assumption the right hand side is positive so we can take logarithms of both sides to obtain

$$H^q \leq \frac{1}{c_3}\left(\log(Cc_2) - \log\left(\sqrt{c_4^2 - S} - T\right)\right)$$

and the result follows. □

If we are unlucky then we just increase the value of the constant $C$ and repeat the algorithm again. If we are continually unlucky then we suppose that this is because one of the following applies:

1. a bound on $H$ is non-existent.
2. a bound on $H$ is very large.
3. the vector $\vec{y}$ actually is a non-zero vector in the lattice $\mathcal{L}$
4. the $\alpha_i$ are linearly dependent over $\mathbb{Q}$.

On the occasions when we do apply this method we have good reasons to believe that $H$ is bounded by a number significantly smaller than $X_0$ and we know that the $\alpha_i$ are linearly dependent over $\mathbb{Q}$.

**VI.3.2. Case 2: $\Re(\alpha_i) = 0$ for all $i$.** In this case we can set $\alpha_i' = \alpha_i/\sqrt{-1}$ and apply the method in Case 1.

**VI.3.3. Case 3: The general case.** We could use the first case here as it is clear either the real or imaginary part of our linear form would satisfy the requisite inequalities. However, for large $H$ we want the real and imaginary parts to be *simultaneously* close to zero. We therefore use the approximation lattices we have already mentioned for simultaneous approximation of a set of linear forms.

Relabel the $\alpha_i$ such that

$$\Re(\alpha_n)\Im(\alpha_{n-1}) - \Re(\alpha_{n-1})\Im(\alpha_n) \neq 0.$$

Then choose a constant $C$ of the order of $X_0^{n/2}$. Define $\mathcal{L}$ to be the lattice generated by the columns of the matrix

$$\mathcal{A} = \begin{pmatrix} 1 & & & & 0 \\ & \ddots & & & \\ 0 & & 1 & & \\ [C\Re(\alpha_1)] & \dots & \dots & [C\Re(\alpha_{n-1})] & [C\Re(\alpha_n)] \\ [C\Im(\alpha_1)] & \dots & \dots & [C\Im(\alpha_{n-1})] & [C\Im(\alpha_n)] \end{pmatrix} \in \mathbb{Z}^{n \times n}.$$

The choice of $C$ is made as before to try and make the size of the first element in an LLL–reduced basis have order $X_0$. Let

$$\vec{y} = \begin{pmatrix} 0 \\ \vdots \\ 0 \\ -[C\Re(\alpha_0)] \\ -[C\Im(\alpha_0)] \end{pmatrix} \in \mathbb{Z}^n.$$

We then apply de Weger's variant of LLL to find a lower bound, $c_4$, on $\ell(\mathcal{L}, \vec{y})$.

LEMMA VI.2. *Set $S = \sum_{i=1}^{n-2} X_i^2$ and $T = (1 + \sum_{i=1}^{n} X_i)/\sqrt{2}$. If $c_4^2 \geq T^2 + S$ then either*

$$H \leq \sqrt[q]{\frac{1}{c_3}\left(\log(Cc_2) - \log\left(\sqrt{c_4^2 - S} - T\right)\right)}$$

*or $x_1 = \cdots = x_{n-2} = 0$ and*

$$\begin{aligned} x_{n-1}[C\Re(\alpha_{n-1})] + x_n[C\Re(\alpha_n)] &= [C\Re(\alpha_0)], \\ x_{n-1}[C\Im(\alpha_{n-1})] + x_n[C\Im(\alpha_n)] &= [C\Im(\alpha_0)]. \end{aligned}$$

PROOF. Define $\Phi_1$ and $\Phi_2$ as follows:

$$\begin{aligned} \Phi_1 &= [C\Re(\alpha_0)] + \sum_{i=1}^{n} x_i \left[C\Re(\alpha_i)\right], \\ \Phi_2 &= [C\Im(\alpha_0)] + \sum_{i=1}^{n} x_i \left[C\Im(\alpha_i)\right]. \end{aligned}$$

Notice that we have

$$|\Phi_1 + \sqrt{-1}\Phi_2 - C(\alpha_0 + \sum_{i=1}^{n} x_i\alpha_i)| \leq (1 + \sum_{i=1}^{n} X_i)/\sqrt{2} = T,$$

and hence

$$|\Phi_1 + \sqrt{-1}\Phi_2| \leq T + Cc_2e^{-c_3H^q}.$$

Now consider the lattice point $\vec{x} = \mathcal{A}\vec{z}$ where

$$\vec{z} = \begin{pmatrix} x_1 \\ \vdots \\ x_n \end{pmatrix}$$

then

$$\vec{x} - \vec{y} = \begin{pmatrix} x_1 \\ \vdots \\ x_{n-2} \\ \Phi_1 \\ \Phi_2 \end{pmatrix}.$$

Hence, either $\vec{x} = \vec{y}$ or

$$\begin{aligned} c_4^2 &\leq \ell(\mathcal{L}, \vec{y}) \leq \sum_{i=1}^{n-2} x_i^2 + \Phi_1^2 + \Phi_2^2, \\ &\leq S + |\Phi_1 + \sqrt{-1}\Phi_2|^2, \\ &\leq S + \left(T + Cc_2 e^{-c_3 H^q}\right)^2. \end{aligned}$$

Now by assumption $c_4^2 - S$ is positive so we have

$$e^{-c_3 H^q} \geq \frac{1}{Cc_2}\left(\sqrt{c_4^2 - S} - T\right)$$

Again by assumption the right hand side is positive so on taking logarithms we obtain the inequality

$$H^q \leq \frac{1}{c_3}\left(\log(Cc_2) - \log\left(\sqrt{c_4^2 - S} - T\right)\right)$$

and the result follows. □

For example suppose that we wish to find all solutions, with $X = \max(|x_i|)$ less than or equal to $X_0 = 10^{30}$, of the following inequality:

$$|x_1 \log 2 + x_2 \log 3 + x_3 \log 5| \leq 2e^{-X}.$$

So in our previous notation we have $c_2 = 2$, $c_3 = 1$ and $X_1 = X_2 = X_3 = 10^{30}$. Our heuristic says choose $C$ to be around $10^{90}$. However, I prefer nice round numbers so we shall use $C = 10^{100}$.

We wish to compute an LLL–reduced basis of the lattice, $\mathcal{L}$, generated by the columns of the matrix

$$\begin{pmatrix} 1 & 0 & 0 \\ 0 & 1 & 0 \\ [C\log 2] & [C\log 3] & [C\log 5] \end{pmatrix}.$$

We find that the first vector in an LLL–reduced basis is given by

$$\begin{pmatrix} -1515246263903680163735468625616799 \\ -502897304507254890263203391695738 \\ 1165937255867757166304329056366403 \end{pmatrix}$$

From Theorem V.9 we find that we can take $c_1 = 1.000$ and then we see that

$$\ell(\mathcal{L}, \vec{0})^2 \geq c_1^{-1}\|\vec{b}_1\|^2 = 0.3908 \cdot 10^{67} = c_4^2.$$

We plug this into Lemma VI.1 to conclude that

$$X \leq 154.$$

Note how small this new bound is in comparison to our original bound of $10^{30}$.

We now repeat the process but with $C = 10^{10}$ and $X_1 = X_2 = X_3 = 154$. The initial basis is now given by the columns of the matrix

$$\begin{pmatrix} 1 & 0 & 0 \\ 0 & 1 & 0 \\ 6931471806 & 10986122887 & 16094379124 \end{pmatrix}.$$

Applying de Weger's LLL–algorithm we find the new LLL basis is given by the columns of the matrix

$$\begin{pmatrix} -399 & -1634 & -2606 \\ 1412 & -1502 & 147 \\ -358 & -1882 & 2681 \end{pmatrix}.$$

Hence we conclude from this second application that a lower bound on $\ell(\mathcal{L}, \vec{0})$ is given by $c_4 = 1485.413$. From which we conclude, from Lemma VI.1, that

$$X \leq 16.$$

We leave it now as an 'easy' exercise to compute which of the $32^3$ remaining possibilities satisfies our original inequality.

## VI.4. $p$-adic analogues

In the last section we looked at when a linear form was extremely small in the sense of

$$|\alpha_0 + \sum_{i=1}^{n} x_i\alpha_i| \leq c_2 e^{-c_3 H^q}$$

where the $\alpha_i \in \mathbb{C}$ are linearly independent over $\mathbb{Q}$. Now the natural generalization to the $p$-adic numbers would be for the $\alpha_i$ to lie in $\mathbb{Q}_p$ or $\mathbb{Q}_p(\theta)$ and the absolute value on the left hand side to be the $p$-adic absolute value. It is this precise generalization which we shall consider in this section.

Firstly we reinterpret the inequality we are given. Suppose $\alpha_i \in \mathbb{Q}_p(\theta)$, some finite extension of $\mathbb{Q}_p$. If we have

$$|\alpha_0 + \sum_{i=1}^{n} x_i\alpha_i|_p \leq c_2 e^{-c_3 H^q}$$

and given that $|.|_p = p^{-\mathrm{ord}_p(.)}$ we find

$$-\mathrm{ord}_p\left(\alpha_0 + \sum_{i=1}^{n} x_i\alpha_i\right)\log p \leq \log c_2 - c_3 H^q.$$

So setting $c_5 = c_3/\log p$ and $c_6 = \log c_2/\log p$ we obtain the inequality

$$\mathrm{ord}_p\left(\alpha_0 + \sum_{i=1}^{n} x_i\alpha_i\right) \geq c_5 H^q - c_6.$$

In other words if $H$ is large then $p$ must divide the linear form to a very high power.

**VI.4.1. A special case.** To get the idea behind what is to follow we first present a very special case; namely we assume $\alpha_i \in \mathbb{Z}_p$. We assume we have large upper bounds $X_i$ on $|x_i|$ and we want to deduce a small upper bound on $H$. Set as before $X_0 = \max X_i$. The constant $C$ in the archimedian case is now replaced by an integer constant $u$ which should be chosen so that $p^u \geq X_0^{n+1}$. The reason for the choice of $u$ of about this size is to have a chance of getting a value of $\|\vec{b}_1\|$ of the required size.

We then let $\alpha^{\{u\}}$, for $\alpha \in \mathbb{Z}_p$, denote the unique rational integer in the interval $[0, \dots, p^u - 1]$ such that

$$\alpha \equiv \alpha^{\{u\}} \pmod{p^u}.$$

Given this definition we define a $p$-adic approximation lattice, for the linear form, to be

$$\mathcal{A} = \begin{pmatrix} 1 & & & 0 \\ & \ddots & & \\ 0 & & 1 & \\ \alpha_1^{\{u\}} & \dots & \alpha_n^{\{u\}} & p^u \end{pmatrix} \in \mathbb{Z}^{(n+1)\times(n+1)}.$$

We also consider the vector

$$\vec{y} = \begin{pmatrix} 0 \\ \vdots \\ 0 \\ -\alpha_0^{\{u\}} \end{pmatrix} \in \mathbb{Z}^{n+1}.$$

Let $\mathcal{L}$ denote the lattice generated over $\mathbb{Z}$ by the columns of the matrix $\mathcal{A}$. By applying de Weger's variant of LLL we can find a lower bound, $c_7$, on $\ell(\mathcal{L}, \vec{y})$. Hopefully this is large enough to bound $H$ using the next lemma. If it is not large enough then we just return to the beginning, make $u$ a little larger and repeat the whole process.

LEMMA VI.3. *If $c_7 > \sqrt{n}X_0$ then either $x_1 = \dots = x_n = 0$ or*

$$H < \sqrt[q]{(u + c_6)/c_5}.$$

PROOF. Assume the result is false i.e. $H \geq \sqrt[q]{(u + c_6)/c_5}$. In which case we have that $p^u$ divides our linear form i.e.

$$\operatorname{ord}_p \left( \alpha_0 + \sum_{i=1}^{n} x_i \alpha_i \right) \geq u.$$

But this also then holds, by definition of $\alpha_i^{\{u\}}$, with $\alpha_i$ replaced by $\alpha_i^{\{u\}}$. Hence

$$\operatorname{ord}_p \left( \alpha_0^{\{u\}} + \sum_{i=1}^{n} x_i \alpha_i^{\{u\}} \right) \geq u.$$

So

$$z = \frac{\alpha_0^{\{u\}} + \sum_{i=1}^n x_i \alpha_i^{\{u\}}}{p^u} \in \mathbb{Z}.$$

If we then consider the lattice point

$$\vec{x} = \mathcal{A} \begin{pmatrix} x_1 \\ \vdots \\ x_n \\ -z \end{pmatrix} = \begin{pmatrix} x_1 \\ \vdots \\ x_n \\ -\alpha_0^{\{u\}} \end{pmatrix},$$

we see that by definition of $\vec{y}$,

$$\vec{x} - \vec{y} = \begin{pmatrix} x_1 \\ \vdots \\ x_n \\ 0 \end{pmatrix}.$$

Then either $\vec{x} = \vec{y}$ or $c_7^2 \leq \sum_{i=1}^n x_i^2 \leq nX_0^2$ which is a contradiction. □

**VI.4.2. The general case.** We now look at the general case where $\alpha_i \in \mathbb{Q}_p(\theta)$, where $\theta$ is a $p$-adic algebraic integer. We first need to reduce the study of one linear form with coefficients in $\mathbb{Q}_p(\theta)$ to the study of a set of linear forms with coefficients in $\mathbb{Q}_p$. This is exactly analogous to how we dealt with linear forms in the archimedian case with non-zero real and imaginary parts. Put $m = [\mathbb{Q}_p(\theta) : \mathbb{Q}_p]$ the degree of the $p$-adic extension defined by $\theta$, which we can assume without loss of generality is a $p$-adic algebraic integer.

LEMMA VI.4. *Write*

$$\alpha_i = \sum_{j=0}^{m-1} \alpha_{i,j} \theta^j$$

*where* $\alpha_{i,j} \in \mathbb{Q}_p$. *If*

$$\operatorname{ord}_p \left( \alpha_0 + \sum_{i=1}^n x_i \alpha_i \right) \geq c_5 H - c_6$$

*then for* $j = 0, \ldots, m-1$ *we have*

$$\operatorname{ord}_p \left( \alpha_{0,j} + \sum_{i=1}^n x_i \alpha_{i,j} \right) \geq c_5 H - c_6 - \frac{1}{2} \operatorname{ord}_p(D(\theta)).$$

*where* $D(\theta)$ *is the discriminant of* $\theta$.

PROOF. There are $m$ $p$-adic conjugates of $\theta$ which we label $\theta^{(i)}$, these correspond as in the archimedian case to the roots of the minimal polynomial over $\mathbb{Q}_p$ of $\theta$. We hence have $m$ conjugate linear forms to consider which we write as

$$\Lambda^{(i)}(\vec{x}) = \sum_{j=0}^{n-1} \Lambda_j(\vec{x}) \theta^{(i)j}$$

where $\Lambda_j(\vec{x})$ is the linear form with $\mathbb{Q}_p$ coefficients given by

$$\Lambda_j(\vec{x}) = \alpha_{0,j} + \sum_{i=1}^{n} x_i\alpha_{i,j}.$$

We hence obtain the matrix equation

$$\begin{pmatrix} \Lambda^{(1)} \\ \vdots \\ \Lambda^{(m)} \end{pmatrix} = \begin{pmatrix} 1 & \theta^{(1)} & \dots & \theta^{(1)m-1} \\ \vdots & \vdots & & \vdots \\ 1 & \theta^{(m)} & \dots & \theta^{(m)m-1} \end{pmatrix} \begin{pmatrix} \Lambda_0 \\ \vdots \\ \Lambda_{m-1} \end{pmatrix} = A \begin{pmatrix} \Lambda_0 \\ \vdots \\ \Lambda_{m-1} \end{pmatrix}.$$

As $\theta$ is a $p$-adic algebraic integer one can write the inverse the matrix $A$ as

$$A^{-1} = \frac{1}{\prod_{1\leq i<j\leq m}(\theta^{(i)} - \theta^{(j)})} \begin{pmatrix} \tau_{1,1} & \dots & \tau_{1,m} \\ \vdots & & \vdots \\ \tau_{m,1} & \dots & \tau_{m,m} \end{pmatrix}$$

where $\tau_{i,j}$ are also $p$-adic algebraic integers, i.e. $\mathrm{ord}_p(\tau_{i,j}) \geq 0$. Now we have the equation

$$\begin{pmatrix} \Lambda_0 \\ \vdots \\ \Lambda_{m-1} \end{pmatrix} = A^{-1} \begin{pmatrix} \Lambda^{(1)} \\ \vdots \\ \Lambda^{(m)} \end{pmatrix}$$

and so, on noting that $\mathrm{ord}_p(\Lambda^{(i)})$ does not depend on $i$, we have

$$\begin{aligned} \mathrm{ord}_p(\Lambda_i) &\geq \mathrm{ord}_p(\Lambda) - \mathrm{ord}_p\left(\prod_{1\leq i<j\leq m}(\theta^{(i)} - \theta^{(j)})\right), \\ &\geq c_5H - c_6 - \frac{1}{2}\mathrm{ord}_p(D(\theta)), \end{aligned}$$

which is what we wanted to prove. □

Having reduced our problem to $m$ linear forms with coefficients in $\mathbb{Q}_p$ we now need to make the coefficients of our linear forms be elements of $\mathbb{Z}_p$. We choose $\lambda \in \mathbb{Q}_p$ such that

$$\mathrm{ord}_p(\lambda) = \min_{1\leq i\leq n}\left(\min_{0\leq j\leq m-1}(\mathrm{ord}_p(\alpha_{i,j})\right) = c_8.$$

We use the next lemma to remove degenerate cases from the discussion.

LEMMA VI.5. *If* $c_8 > \mathrm{ord}_p(\alpha_{0,i})$ *for some* $i$ *then we have*

$$H < \frac{1}{c_5}\left(c_6 + c_8 + \frac{1}{2}\mathrm{ord}_p(D(\theta))\right).$$

PROOF. For the $i$ in question, such that the $p$-adic order of $\alpha_{0,i}$ is less than $c_8$, we find

$$\mathrm{ord}_p(\Lambda_i) = \mathrm{ord}_p(\alpha_{0,i}) < c_8.$$

Hence

$$c_8 > \mathrm{ord}_p(\Lambda_i) \geq c_5H - c_6 - \frac{1}{2}\mathrm{ord}_p(D(\theta)),$$

from which the stated result follows. □

Hence we can assume that for all $i$ we have $c_8 \leq \text{ord}_p(\alpha_{0,i})$. In particular this means that for all $i$ and $j$ we have

$$\beta_{i,j} = \alpha_{i,j}/\lambda \in \mathbb{Z}_p.$$

So the linear forms given by $\Lambda_i/\lambda$ have $\mathbb{Z}_p$ coefficients. If we then write

$$\Delta_j = \beta_{0,j} + \sum_{i=1}^{n} x_i \beta_{i,j}$$

we find that

$$\text{ord}_p(\Delta_j) = \text{ord}_p(\Lambda_j/\lambda) \geq c_5 H - c_6 - \frac{1}{2}\text{ord}_p(D(\theta)) - c_8 = c_5 H - c_9.$$

So we have now reduced the problem to a set of $m$ linear forms with coefficients in $\mathbb{Z}_p$. We could at this stage just select one of these linear forms and then apply the method of the special case mentioned above. However, this is not really what we want. When $H$ is large the above inequality tells us that the $\Delta_j$ are *simultaneously* very close to zero in the $p$-adic metric. We therefore wish to *simultaneously* approximate the $m$ linear forms that we have. It has already been alluded to how to do this for real linear forms. We now give the obvious extension to $p$-adic linear forms.

We retain all the previous notation. Choose $u \in \mathbb{N}$ such that $p^u$ is about the size of $X_0^{1+n/m}$. We then consider the square $n+m$ dimensional matrix:

$$\mathcal{A} = \begin{pmatrix} 1 & & & & 0 \\ & \ddots & & & \\ 0 & & 1 & & \\ \beta_{1,0}^{\{u\}} & \cdots & \beta_{n,0}^{\{u\}} & p^u & & 0 \\ \vdots & & \vdots & & \ddots & \\ \beta_{1,m-1}^{\{u\}} & \cdots & \beta_{n,m-1}^{\{u\}} & 0 & & p^u \end{pmatrix}$$

and let $\mathcal{L}$ denote the lattice generated over $\mathbb{Z}$ by the columns of $\mathcal{A}$. We shall also require the vector

$$\vec{y} = \begin{pmatrix} 0 \\ \vdots \\ 0 \\ -\beta_{0,0}^{\{u\}} \\ \vdots \\ -\beta_{0,m-1}^{\{u\}} \end{pmatrix} \in \mathbb{Z}^{n+m}.$$

We find a lower bound, $c_{10}$, on $\ell(\mathcal{L}, \vec{y})$ using de Weger's LLL; then hopefully we can deduce a bound on $H$ using the following lemma.

LEMMA VI.6. *If* $c_{10} > \sqrt{n}X_0$ *then either* $x_1 = \cdots = x_n = 0$ *or*

$$H < (u + c_9)/c_5.$$

PROOF. As in the special case considered earlier we assume the contrary, that $H \geq (u + c_9)/c_5$. But then

$$\mathrm{ord}_p(\Delta_j) = c_5 H - c_9 \geq u.$$

Then it follows that for all $j$ we have

$$\mathrm{ord}_p\left(\beta_{0,j}^{\{u\}} + \sum_{i=1}^{n} x_i \beta_{i,j}^{\{u\}}\right) \geq u,$$

which in turn means that for all $j$,

$$z_j = \frac{\beta_{0,j}^{\{u\}} + \sum_{i=1}^{n} x_i \beta_{i,j}^{\{u\}}}{p^u} \in \mathbb{Z}.$$

If we then consider the lattice point

$$\vec{x} = \mathcal{A}\begin{pmatrix} x_1 \\ \vdots \\ x_n \\ -z_0 \\ \vdots \\ -z_{m-1} \end{pmatrix} = \begin{pmatrix} x_1 \\ \vdots \\ x_n \\ -\beta_{0,0}^{\{u\}} \\ \vdots \\ -\beta_{0,m-1}^{\{u\}} \end{pmatrix},$$

we find

$$\vec{x} - \vec{y} = \begin{pmatrix} x_1 \\ \vdots \\ x_n \\ 0 \\ \vdots \\ 0 \end{pmatrix}.$$

But then we see that either $\vec{x} = \vec{y}$ or

$$c_{10}^2 \leq \ell(\mathcal{L}, \vec{y})^2 \leq \sum_{i=1}^{n} x_i^2 \leq nX_0^2,$$

which is a contradiction. □

If the above lemma does not give us a bound on $H$ then we simply increase the value of $u$ and start again.

The astute will have realized that this may cause a computational problem using an $m + n$ square matrix as opposed to an $n + 1$ square matrix, which would be the case if we just selected one linear form to work with. However, by choosing an $m + n$ dimensional square matrix, the expected value of $u$ needed will be much smaller. Hence the entries in the matrix which we need to perform LLL on will be smaller. In practice the advantage in the above method is that the bounds produced are generally much better than using only one linear form.

## VI.5. Exercises

1). Compute all integer solutions to the inequality

$$\begin{aligned} |x\log 2 + y\log 3 + z\log 5 - \log 7| &\leq e^{-X}, \\ X = \max(|x|,|y|,|z|) &\leq 10^{30}. \end{aligned}$$

2). Find an upper bound on $H$ from the following simultaneous system of linear inequalities:

$$\begin{aligned} |x\log 2 + y\log 3 + z\log 5| &\leq 2e^{-H}, \\ |x\log 3 + y\log 2 + z\log 5| &\leq 2e^{-H}, \\ |x\log 5 + y\log 2 + z\log 3| &\leq 2e^{-H}, \end{aligned}$$

where $(x, y, z) \in \mathbb{Z}^3$ satisfy $\max(|x|,|y|,|z|) \leq 10^{30}$.

3). Choose a prime $p$ greater than 5. Determine an upper bound on $H$ from the inequality

$$\mathrm{ord}_p(x_1 \log_p 2 + x_2 \log_p 3 + x_3 \log_p 5) \geq 8H - 3,$$

where $(x, y, z) \in \mathbb{Z}^3$ satisfy $\max(|x|,|y|,|z|) \leq 10^{30}$.

# Part 2

# Methods using linear forms in logarithms

CHAPTER VII

# Thue equations

In this chapter we examine the method of Tzanakis and de Weger for Thue equations. In later chapters the method will be generalized to other types of diophantine equations such as Thue–Mahler equations and discriminant form equations. What is interesting about the following algorithm is that it was the first practical generic method to solve a wide class of diophantine equations. Much theoretical work had been done on bounding the solutions to such equations, but until the advent of Tzanakis and de Weger's algorithm it was not possible to produce a general algorithm, which could be applied in practice to equations of interest. There were, however, a variety of ad hoc techniques such as those to be found in [**138**, Chapter 23], [**49**], [**188**], [**200**], [**199**] and [**190**]. These either used special properties of the equations or were Skolem's method in disguise.

Some authors, see [**186**] and [**149**], had previously used the LLL–algorithm to solve specific problems or special cases. However, Tzanakis and de Weger were the first to describe the method in complete generality. There is a good survey of the start of the art up to around 1988 by A. Pethő in [**144**].

We shall end this chapter by showing how we can use the method to find solutions to Thue equations to find all the integral points on an elliptic curve. This will be the first of three such methods to find integral points on elliptic curves. Although the method in this chapter is not the most efficient it is practical. The procedure is reminiscent of the method of 2-descent which we shall come across later on in Chapter XII.

## VII.1. Thue equations

Recall, a Thue equation is a diophantine equation of the form

$$F(X,Y)=m, \tag{VII.1}$$

where $F(X,Y)\in\mathbb{Z}[X,Y]$ is a binary form of degree $n\geq 3$ and $m$ is a fixed integer. We are interested in solutions $(X,Y)\in\mathbb{Z}^2$. For ease of exposition we shall assume that $F(X,1)$ is a monic polynomial, the adaptations for the general case are trivial and are left as an exercise. In addition, we note that the case of reducible forms follows trivially from the case for irreducible forms, so we shall assume in describing the method that $F(X,Y)$ is irreducible.

By a result of Thue [**196**] dating back to 1909, we know that there are only finitely many solutions to such a problem. However, it was not until the ground breaking work of Baker in the late 1960s [**4**] that a theoretical algorithm could be given to find the solutions. In the 1970s and the early 1980s many such equations were solved by applying Baker's method with a variety of computational techniques to find all the solutions in a reasonably efficient way. In 1989 Tzanakis and de Weger [**202**] showed that combining the computational diophantine approximation techniques in VI.3 with Baker's theory could give a general, practical algorithm to solve such equations.

We shall first dispose of a rather simple case.

LEMMA VII.1. *Suppose $F(X,1)$ has no real roots, then the solutions to equation (VII.1) satisfy*

$$|Y| \leq \frac{|m|}{\min_{1\leq i\leq n} |\Im(\theta^{(i)})|},$$

*where $\theta$ denotes a root of $F(X,1)$.*

PROOF. Suppose $(X,Y)$ is a solution and $\theta^{(i)}$ is chosen to make the value of $|X-\theta^{(i)}Y|$ less than $|m|$. We then have that

$$|\Im(\theta^{(i)})Y| \leq |X-\theta^{(i)}Y| \leq |m|,$$

from which the result follows. □

Hence if $F(X,1)$ has no real roots then we can easily solve our equation. We shall therefore assume that $F(X,1)$ has $s$ real roots (with $s \geq 1$) and $t$ complex conjugate roots, so $n = s+2t$. Let $\theta$ denote a root of $F(X,1)$ and $K = \mathbb{Q}(\theta)$. We order the conjugates of $\theta$ in the standard way so that

$$\begin{aligned} \theta^{(i)} &\in \mathbb{R} && \text{if } 1 \leq i \leq s, \\ \theta^{(i)} &= \overline{\theta^{(i+t)}} && \text{if } s+1 \leq i \leq s+t. \end{aligned}$$

We set

$$\beta^{(i)} = X - \theta^{(i)}Y.$$

Hence, by the unique factorization of the principal ideal $(X-\theta Y)$,

$$\beta^{(i)} = \mu^{(i)}\epsilon^{(i)},$$

where $\epsilon$ is a unit of $\mathcal{O}_K$ and $\mu$ comes from a complete set of representatives of the elements of $\mathcal{O}_K$ of norm $m$ up to associativity.

By appealing to our favorite number theory computer package we hope we can compute the $\mu$ and a set of generators (fundamental units) for the unit group of $\mathcal{O}_K$. In practice this always appears to be possible using the latest algorithms [**32**] and [**34**], unless the degree $n$ is too large or the regulator and/or class number of $K$ is too big. This is, of course, assuming one is willing to believe certain conjectures such as a generalized Riemann hypothesis (GRH). If one is not willing to assume such rash conjectures then there are slower non-conjectural algorithms that one can use, see [**155**] and [**154**].

So we assume that we can compute a complete set of representatives for the $\mu$, a set of fundamental units $\eta_1, \ldots, \eta_r$, where $r = s + t - 1$, and all possible units of finite order $\xi$. Hence we can write $\beta$ as

$$\beta = \xi\mu \prod_{j=1}^{r} \eta_j^{a_i},$$

for some unknown integers $a_i$. We now take three distinct indices labelled $i, j, k$. The indices $j$ and $k$ are arbitrary but the index $i$ will be defined by

$$|\beta^{(i)}| = \min_{1\le l\le n} |\beta^{(l)}|.$$

We do not know *a priori* the value of $i$, so we shall have to perform the following steps for all possible values of $i$. Later we shall show that we can restrict $i$ to correspond to the real embeddings of $K$. From the three chosen indices we obtain three linear equations to solve for $X$ and $Y$,

$$\begin{aligned} X - \theta^{(i)}Y &= \beta^{(i)}, \\ X - \theta^{(j)}Y &= \beta^{(j)}, \\ X - \theta^{(k)}Y &= \beta^{(k)}. \end{aligned}$$

We can solve the first two for $X$ and $Y$ and then substitute the resulting expressions into the third equation to find an identity which the $\beta^{(i)}$ must satisfy to produce a solution to our Thue equation, namely,

$$\beta^{(i)}(\theta^{(j)} - \theta^{(k)}) + \beta^{(j)}(\theta^{(k)} - \theta^{(i)}) + \beta^{(k)}(\theta^{(i)} - \theta^{(j)}) = 0.$$

This identity we have met before when considering Skolem's method in Section III.2. Thus we obtain the following equation:

$$\alpha_1\tau_1 + \alpha_2\tau_2 + 1 = 0, \tag{VII.2}$$

where we have set

$$\alpha_1 = \frac{\xi^{(i)}\mu^{(i)}(\theta^{(j)} - \theta^{(k)})}{\xi^{(j)}\mu^{(j)}(\theta^{(k)} - \theta^{(i)})}, \quad \alpha_2 = \frac{\xi^{(k)}\mu^{(k)}(\theta^{(i)} - \theta^{(j)})}{\xi^{(j)}\mu^{(j)}(\theta^{(k)} - \theta^{(i)})}, \tag{VII.3}$$

$$\tau_1 = \prod_{l=1}^{r} \left(\frac{\eta_l^{(i)}}{\eta_l^{(j)}}\right)^{a_l}, \quad \tau_2 = \prod_{l=1}^{r} \left(\frac{\eta_l^{(k)}}{\eta_l^{(j)}}\right)^{a_l}.$$

Now notice that the $\alpha_i$ are fixed and the $\tau_i$ range over a finitely generated subgroup of $L^*$, where $L = K^{Gal}$. Such an equation (VII.2) is called a two term $S$-unit equation. By a result of Siegel there are only finitely many solutions to such an equation. We shall return to such equations in a later chapter as we try to generalize the current method.

Now suppose that $|\alpha_1\tau_1|$ is very, very small; then $-\alpha_2\tau_2$ would have to be very close to 1. But that would mean

$$\Lambda = \log(-\alpha_2\tau_2)$$

would also be very small, where we take the principal value of the logarithm. But $\Lambda$ can also be written as

$$\Lambda = \log(-\alpha_2) + \sum_{l=1}^{r} a_l \log\left(\frac{\eta_l^{(k)}}{\eta_l^{(j)}}\right) + a_0 2\pi\sqrt{-1}$$

for some $a_0 \in \mathbb{Z}$. Hence we are saying that if $\alpha_1\tau_1$ is very small then the linear form in logarithms, $\Lambda$, will also be very small. However, such a linear form in logarithms cannot get arbitrarily close to zero by the celebrated results of Baker, unless the $a_i$ are very large or the linear form is equal to zero. A small change (which we leave as an exercise) will need to be made to our method if our logarithms are not linearly independent over $\mathbb{Q}$. From now on we shall assume that we do not have $\Lambda = 0$. We have hence reduced our problem of solving a diophantine equation to the problem of quantifying how small all the linear forms in logarithms can become.

It will become convenient to set $A = \max_{i\neq 0}(|a_i|)$. Hence if we can find a small bound on $A$ then we can loop through all possible $a_i$ and solve our problem. We note that

$$\begin{aligned} |a_0 2\pi| &= \left|\arg(-\alpha_2\tau_2) - \arg(-\alpha_2) - \sum_{i=1}^{r} a_i \arg\left(\frac{\eta_l^{(k)}}{\eta_l^{(j)}}\right)\right|, \\ &\leq \pi\left(2 + rA\right), \end{aligned}$$

hence if $A \geq 2$ then $|a_0| \leq (r+1)A/2$.

We find from the equation

$$|Y|\ |\theta^{(i)} - \theta^{(j)}| = |\beta^{(i)} - \beta^{(j)}| \leq 2|\beta^{(j)}|$$

that for all $j \neq i$ we have

$$|\beta^{(j)}|^{-1} \leq 2\left(\min_{l\neq m} |\theta^{(m)} - \theta^{(l)}|\right)^{-1} = c_1.$$

On setting

$$c_2 = \max_{l_1\neq l_2\neq l_3\neq l_1} \left|\frac{(\theta^{(l_2)} - \theta^{(l_3)})}{(\theta^{(l_3)} - \theta^{(l_1)})}\right|$$

we find that

$$|\alpha_1\tau_1| \leq c_1 c_2 |\beta^{(i)}| = c_3|\beta^{(i)}|.$$

Hence we require $\beta^{(i)}$ to be small to make $\alpha_1\tau_1$ small and so to make the linear form in logarithms, $\Lambda$, small.

To continue we shall need the following general lemma. This lemma relates the size of the logarithm of a particular conjugate of a unit to the maximum exponent of the chosen fundamental units in the representation of the unit and a kind of 'regulator'. In a later chapter we shall generalize the following to arbitrary finitely generated groups of algebraic numbers using the logarithmic embedding of the $S$-units.

LEMMA VII.2. *Let $K$ be a number field with $r$ fundamental units $\eta_i \in K$ and let $a_i \in \mathbb{Z}$ for $1 \leq i \leq r$. Define $A = \max|a_i|$ and*

$$\epsilon = \prod_{i=1}^{r} \eta_i^{a_i}.$$

*Let $I = \{i_1, \dots, i_r\} \subset \{1, \dots, s+t\}$ denote any set of $r$ distinct indices then following matrix is invertible:*

$$U_I = \begin{pmatrix} \log|\eta_1^{(i_1)}| & \dots & \log|\eta_r^{(i_1)}| \\ \vdots & & \vdots \\ \log|\eta_1^{(i_r)}| & \dots & \log|\eta_r^{(i_r)}| \end{pmatrix}$$

*and there exists an index $t \in I$ such that*

$$|\log|\epsilon^{(t)}|| = \max_{1 \leq l \leq r+1} |\log|\epsilon^{(l)}||$$

*then*

$$|\log|\epsilon^{(t)}|| \geq A/\|U_I^{-1}\|_\infty,$$

*where $\|.\|_\infty$ denotes the infinity norm of a matrix, in other words the row sum norm.*

PROOF. Consider the matrix equation

$$\begin{pmatrix} a_1 \\ \vdots \\ a_r \end{pmatrix} = U_I^{-1} \begin{pmatrix} \log|\epsilon^{(i_1)}| \\ \vdots \\ \log|\epsilon^{(i_r)}| \end{pmatrix}.$$

Taking the infinity matrix norm of both sides we obtain

$$A \leq \|U_I^{-1}\|_\infty \, |\log|\epsilon^{(t)}||,$$

and we are done. □

Applying this lemma in our situation we find that there is an index $t$ such that there is a computable constant $c_4$ which satisfies

$$|\log|\epsilon^{(t)}|| \geq c_4 A.$$

We can take $c_4$ to be the maximum of $\|U_I^{-1}\|_\infty$ over all possible subsets $I$ in the above lemma. The exact value of $t$ is in fact irrelevant all we need know is that it exists. We then choose the constant $c_5$ to be any positive real number with $c_5 < c_4/(n-1)$. We divide our discussion into two cases: either $|\beta^{(i)}| > e^{-c_5 A}$ or $|\beta^{(i)}| \leq e^{-c_5 A}$. The first of these we further divide into two sub cases: according to the lemma above we have either

$$|\epsilon^{(t)}| \geq e^{c_4 A}$$

or

$$|\epsilon^{(t)}| \leq e^{-c_4 A}.$$

**Case A :** $|\beta^{(i)}| > e^{-c_5 A}$ **and** $|\epsilon^{(t)}| \geq e^{c_4 A}$. Firstly we notice we have the inequality

$$|\beta^{(t)}| < |m| \prod_{l \neq t} |\beta^{(l)}|^{-1} < |m||\beta^{(i)}|^{-(n-1)} < |m|e^{(n-1)c_5 A}.$$

So we obtain the inequality

$$e^{c_4 A} \leq |\epsilon^{(t)}| = |\beta^{(t)}|/|\mu^{(t)}| \leq c_6 |m| e^{(n-1)c_5 A},$$

where $c_6 = \max_{1 \leq l \leq n} |\mu^{(l)}|^{-1}$, and so $(c_4 - (n-1)c_5)A \leq \log(c_6|m|)$, which means, in this case, that we have

$$A \leq \frac{\log(c_6|m|)}{c_4 - (n-1)c_5} = A_1.$$

**Case B :** $|\beta^{(i)}| > e^{-c_5 A}$ **and** $|\epsilon^{(t)}| \leq e^{-c_4 A}$. Here we have the inequality

$$e^{-c_4 A} \geq |\epsilon^{(t)}| = |\beta^{(t)}|/|\mu^{(t)}| \geq c_7 e^{-c_5 A},$$

where $c_7 = \min_{1 \leq l \leq n} |\mu^{(l)}|^{-1}$. Hence we obtain the following bound on $A$:

$$A \leq \frac{\log(c_7)}{c_4 - c_5} = A_2.$$

Hence in both cases A and B we have a (hopefully) rather small upper bound on $A$. This is exactly what we wanted to solve our Thue equation. So we can conclude that if $\beta^{(i)}$ is 'large' then we can easily solve our diophantine equation. The problem is that $\beta^{(i)}$ could be very small. It was indeed chosen to be the smallest conjugate of $\beta$ after all. It is to cope with this last case that we will need to use the heavy theory of linear forms in logarithms and the techniques of lattice reduction which we have already introduced.

**Case C :** $|\beta^{(i)}| \leq e^{-c_5 A}$. In this case we find that we have

$$|\alpha_1 \tau_1| \leq c_3 e^{-c_5 A}.$$

We note that for any complex number $z$ if $|z-1| \leq \frac{1}{2}$ then $|\log z| \leq 2|z-1|$ (see Appendix B, Lemma B.2). Thus if

$$A \geq \log(2c_3)/c_5 = A_3$$

then

$$|e^{\Lambda} - 1| = |\alpha_1 \tau_1| \leq 1/2.$$

Hence we find that (as $\Lambda$ is the principal value of a logarithm)

$$|\Lambda| \leq 2c_3 e^{-c_5 A}. \tag{VII.4}$$

We remind ourselves that $\Lambda$ is a linear form in logarithms of algebraic numbers, hence we can apply the theory of such objects, see Appendix A.1, to determine a constant $c_8$ such that if $A \geq 3$

$$\log|\Lambda| > -c_8 \log((r+1)A/2)$$

(provided $\Lambda \neq 0$ which we have already pointed out can be circumvented). Therefore putting the last two inequalities together we find that

$$A < \frac{1}{c_5}\left(c_8 \log A + \log(2c_3) + c_8 \log((r+1)/2)\right).$$

By the lemma of Pethő and de Weger, see Appendix B.1, we obtain an upper bound, $A_4$, on $A$:

$$A \leq A_4 = \frac{2}{c_5}\left(\log(2c_3) + c_8 \log((r+1)/2) + c_8 \log\left(c_8/c_5\right)\right).$$

The trouble is that owing to the large value of the constant $c_8$ the upper bound, $A_4$, in this third case can be rather large. Luckily we also have equation (VII.4), hence using the techniques of Section VI.3, we can reduce this upper bound $A_4$ to something more manageable. Let $A_5$ denote this new reduced upper bound.

We now have (hopefully) a very small overall upper bound on the number $A$ given by $A_6 = \max(3, A_1, A_2, A_3, A_5)$. To solve our equation we need therefore only loop on all possible values of the $a_i$ in the equation

$$X - \theta Y = \mu\xi \prod_{i=1}^{r} \eta_i^{a_i}$$

and equate the coefficients of $\theta^i$, for $i \geq 2$, on the right hand side to zero. This may be a sensible suggestion for small values of $A^r$ but for larger values this can be slightly silly. To simplify this search there are many techniques. One way to proceed is to use the congruence conditions on the exponents, $a_i$, which are derived as a by-product of Skolem's method. Indeed if one applies Skolem's method for many prime numbers one will obtain a set of congruence conditions on the exponents which will hopefully eliminate all exponents which are not actual solutions. This last idea is what we shall develop later into a sieving strategy for general $S$-unit equations.

When performing the above procedure for finding the upper bound on $A$ one has to loop through all possible values of $i \in \{1, \ldots, n\}$; this is rather wasteful as we shall now show. We shall reduce to only looping on $i \in \{1, \ldots, s\}$. To do this we first need to link the size of $|\beta^{(i)}|$ to the size of $|Y|$.

LEMMA VII.3. *If we set*

$$c_9 = \max_{1 \leq i \leq n}\left(\frac{2^{n-1}|m|}{|\frac{\partial F}{\partial X}(\theta^{(i)}, 1)|}\right)$$

*then*

$$|\beta^{(i)}| \leq c_9 |Y|^{-(n-1)}.$$

PROOF. Choose $i$ as above, so that $|\beta^{(i)}| = \min|\beta^{(l)}|$, and consider the equation

$$\prod_{l=1}^{n} |\beta^{(l)}| = |m|.$$

By choice of $i$ we have the following inequality for all values of $l \neq i$:

$$\begin{aligned} |\beta^{(l)}| &\geq \frac{1}{2}\left(|\beta^{(l)}| + |\beta^{(i)}|\right) \\ &\geq \frac{1}{2}|\beta^{(i)} - \beta^{(l)}| \\ &= \frac{1}{2}|Y|\,|\theta^{(i)} - \theta^{(l)}|, \end{aligned}$$

and so

$$\begin{aligned} |\beta^{(i)}| &= |m| \prod_{l \neq i} |\beta^{(l)}|^{-1} \\ &\leq |m| \prod_{l \neq i} \left(\frac{1}{2}|Y|\,|\theta^{(i)} - \theta^{(l)}|\right)^{-1} \\ &= \left(\frac{2^{n-1}|m|}{|\frac{\partial F}{\partial X}(\theta^{(i)}, 1)|}\right) |Y|^{-(n-1)}, \end{aligned}$$

which is what we wanted to show. □

We can now show that we do not need to loop through all possible values of the index $i$ in the main algorithm above. We only need loop through the values of $i$ which correspond to real embeddings of $K$, as long as we assume that $|Y|$ is large enough, which is no real restriction anyway.

LEMMA VII.4. *If we define the constant* $Y_1$ *by*

$$Y_1 = \begin{cases} \left(\frac{c_9}{\min_{s<l\leq n} |\Im(\theta^{(l)})|}\right)^{1/n} & \text{if } t \geq 1, \\ 1 & \text{if } t = 0 \end{cases}$$

*then* $|Y| \geq Y_1$ *implies that* $i$ *must belong to the set* $\{1, \ldots, s\}$.

PROOF. We can clearly assume that $t \geq 1$, otherwise the result is rather trivial. Notice that for any complex number $\alpha$, with non-zero real part, we must have

$$|\Im(\alpha)| < |\alpha|.$$

Suppose $i > s$ and $|Y| \geq Y_1$ then we have, using Lemma VII.3,

$$\begin{aligned} |\Im(\theta^{(i)})| &< |\frac{X}{Y} - \theta^{(i)}| = \frac{|\beta^{(i)}|}{|Y|} \leq c_9|Y|^{-n} \\ &\leq \left(\frac{Y_1}{|Y|}\right)^n \min_{s<l\leq n} |\Im(\theta^{(l)})| \\ &\leq \min_{s<l\leq n} |\Im(\theta^{(l)})| \end{aligned}$$

which is a contradiction. □

For various optimizations of the above method, for instance using continued fractions to find the 'small' solutions, you should consult the original paper by Tzanakis and de Weger [**202**] and the paper by A. Pethő [**146**]. Also see later where we discuss the method of Bilu and Hanrot.

There has been a lot of work on trying to solve families of Thue equations given by a parametrized equation in one or two variables. There have been many papers on this in the last few years as techniques have become more developed: [**195**], [**130**], [**147**], [**132**]. These papers have considered quartic equations with one parameter. Recently these techniques have been extended further; for example in [**150**] a quartic family is considered which depends on two parameters, while in [**99**] a quintic family is considered.

Bombieri and Schmidt have in addition shown the following result

THEOREM VII.5 ([**16**]). *Let $t$ denote the number of prime factors of $m$ and $n$ denote the degree of $F(X,Y)$. Then the equation*

$$|F(X,Y)| = m$$

*has at most*

$$c\, n^{t+1}$$

*solutions, for some positive constant $c$.*

## VII.2. $X^4 - 2Y^4 = \pm 1$

We end our discussion of Thue equations by returning to the example of Section III.2, namely

$$X^4 - 2Y^4 = \pm 1.$$

In an exercise you should have shown that this had exactly six solutions. We shall now show this without the need to make 'lucky' choices for the primes to use in Skolem's method. We recap that in the field $K = \mathbb{Q}(\theta)$ where $\theta^4 - 2 = 0$ we have two fundamental units which we can take to be

$$\eta_1 = 1 + \theta^2 \ , \ \eta_2 = 1 + \theta.$$

Hence it remains to find all possible exponents $a_i$ such that

$$X - \theta Y = \beta = \pm \eta_1^{a_1} \eta_2^{a_2}.$$

We label the roots of $X^4 - 2$ so that

$$\theta^{(1)} \approx -1.189207 \ , \ \theta^{(2)} \approx 1.189207 \ , \ \theta^{(3)} = \overline{\theta^{(4)}} \approx -1.189207\sqrt{-1},$$

from which we can easily calculate the values of $c_1, c_2, c_3$ and $c_4$ above:

$$c_1 = 1.18921 \ , \ c_2 = 1.4143 \ , \ c_3 = 1.6819 \ , \ c_4 = 1.9515.$$

We then choose $c_5 = 0.65$, which certainly satisfies the inequality $c_5 < c_4/(n-1) = c_4/3$. It is easy to see that the constants $c_6$ and $c_7$ in this case are both equal to 1, from which we can conclude that cases A and B provide the trivial upper bounds of $A_1 = A_2 = 0$.

We now need to split into 2 cases (by Lemma VII.4) for if $Y \geq Y_1 = 1$ there are two choices (namely $i = 1$ and $i = 2$) for the index $i$ such that

$$|\beta^{(i)}| = \min_{1 \leq l \leq 4} |\beta^{(l)}|.$$

**Case** $i = 1$. We need to make a choice of $j$ and $k$, which we make as $j = 3$ and $k = 2$. This is not the choice one would make if one followed letter by letter the algorithm of Tzanakis and de Weger. (They would choose $j = 3$, $k = 4$.) The reason we prefer this choice is to make the linear forms in the logarithms non-trivial and to make sure we have a linear form in logarithms with non-trivial real and imaginary parts. This is because the LLL–reduction considered in Section VI.3 seems to work better when one is reducing a linear form with both a non-trivial real and imaginary part.

We now need to look at all the values in our linear form in logarithms,

$$\Lambda = \log(-\alpha_2) + \sum_{l=1}^{r} a_l \log\left(\frac{\eta_l^{(2)}}{\eta_l^{(3)}}\right) + a_0 2\pi\sqrt{-1},$$

for some $a_0 \in \mathbb{Z}$. Firstly we notice that

$$\alpha_2 = \pm\left(\frac{\theta^{(1)} - \theta^{(3)}}{\theta^{(2)} - \theta^{(1)}}\right) = \pm(1 - \sqrt{-1})/2;$$

we also have

$$\left(\frac{\eta_1^{(2)}}{\eta_1^{(3)}}\right) = \left(\frac{1+\theta^2}{1-\theta^2}\right), \quad \left(\frac{\eta_2^{(2)}}{\eta_2^{(3)}}\right) = \left(\frac{1-\theta}{1+\sqrt{-1}\theta}\right).$$

The first of this pair has minimal polynomial $X^2 + 6X + 1$ while the second has minimal polynomial

$$X^8 + 8X^7 + 44X^6 - 136X^5 + 230X^4 - 136X^3 + 44X^2 + 8X + 1.$$

It is then an easy matter to compute

$$h(\alpha_2) = 0.3465\ , \ h\left(\frac{\eta_1^{(2)}}{\eta_1^{(3)}}\right) = 0.88137\ , \ h\left(\frac{\eta_2^{(2)}}{\eta_2^{(3)}}\right) = 0.61211.$$

Then the modified height, in the sense of Appendix A.1, of the three algebraic numbers is then

$$\begin{aligned} h_m(\alpha_2) &= \max\{0.3465, 0.1073, 0.125\} = 0.3465, \\ h_m\left(\frac{\eta_1^{(2)}}{\eta_1^{(3)}}\right) &= \max\{0.88137, 0.4503, 0.125\} = 0.88137, \\ h_m\left(\frac{\eta_2^{(2)}}{\eta_2^{(3)}}\right) &= \max\{0.61211, 0.1171, 0.125\} = 0.61211. \end{aligned}$$

Hence we can give a lower bound for the linear form $|\Lambda|$, using the theorem in Appendix A.1,

$$\log|\Lambda| \geq -c_8 \log(3A/2),$$

where

$$c_8 = 18 \cdot 5! \cdot 4^5 \cdot 256^6 (\log 64) h_m(-1) h_m(\alpha_2) h_m\left(\frac{\eta_1^{(2)}}{\eta_1^{(3)}}\right) h_m\left(\frac{\eta_2^{(2)}}{\eta_2^{(3)}}\right) = 1.902 \cdot 10^{20}.$$

Hence we can conclude that in Case C (with $i = 1$) we have that if $A \geq A_3 = 1.866$ then

$$A \leq A_4 = 2.8 \cdot 10^{22}.$$

Using this upper bound with the inequality (VII.4) we can apply the techniques of VI.3 to try to deduce a smaller bound on $A$.

Firstly, setting $C = 10^{40}$ in the notation of Section VI.3, we need to find an LLL–reduced matrix for the lattice generated by the columns of the matrix

$$\begin{pmatrix} 1 & 0 & 0 \\ 1762747174039086050465218649959584618O563 & 3428526368194948697700456318381422611295 & 0 \\ 3141592653589793238462643383279502884197 2 & 8716111622538728427083539842926495267420 & [10^{40} 2\pi] \end{pmatrix}$$

We find, using de Weger's algorithm, that an LLL–reduced basis is given by the columns of the matrix

$$\begin{pmatrix} -473370331232103025809742217 & -435148611645581868271795O8 & 446245042641595366633146255 \\ 153886283708763213148976734 & 481264526108538228029397736 & -138948196427026090773798155 \\ -98044060959630723435611936 & 310551489586956203244388693 & 687498403802257053617844755 \end{pmatrix}.$$

If we set $\vec{y} = (0, \pm[C/2], \mp[C/2])^t$ then we find that $\ell(\mathcal{L}, \vec{y}) \geq 9.93624 \cdot 10^{52}$ by Theorem V.10. This leads us to deduce, by Lemma VI.2, that

$$A \leq 49.$$

We now repeat the process again, this time using $C = 10^5$. Now we wish to find an LLL–reduced basis of the new lattice generated by the columns of

$$\begin{pmatrix} 1 & 0 & 0 \\ 176275 & 34285 & 0 \\ 314159 & 87161 & 628319 \end{pmatrix}.$$

We find that an LLL–reduced basis of this lattice is given by the columns of

$$\begin{pmatrix} -1965 & -544 & -2284 \\ 980 & 1545 & -3345 \\ -513 & 1817 & 1114 \end{pmatrix}.$$

Applying Theorem V.10 with the vector $\vec{y} = (0, \pm[C/2], \mp[C/2])^t$ we find $\ell(\mathcal{L}, \vec{y}) \geq 1301689.087$ and a new bound of $A \leq 8$.

**Case** $i = 2$**.** We again need to make a choice of $j$ and $k$, which we make as $j = 3$ and $k = 1$. We now look at all the values in the linear form

$$\Lambda = \log(-\alpha_2) + \sum_{l=1}^{r} a_l \log\left(\frac{\eta_l^{(1)}}{\eta_l^{(3)}}\right) + a_0 2\pi\sqrt{-1}$$

for some $a_0 \in \mathbb{Z}$. Firstly we notice that

$$\alpha_2 = \pm\left(\frac{\theta^{(2)} - \theta^{(3)}}{\theta^{(1)} - \theta^{(2)}}\right) = \pm(1+\sqrt{-1})/2,$$

we also have

$$\left(\frac{\eta_1^{(1)}}{\eta_1^{(3)}}\right) = \left(\frac{1+\theta^2}{1-\theta^2}\right), \quad \left(\frac{\eta_2^{(1)}}{\eta_2^{(3)}}\right) = \left(\frac{1+\theta}{1+\sqrt{-1}\theta}\right).$$

These last two have the same minimal polynomials as before namely $X^2 + 6X + 1$ and

$$X^8 + 8X^7 + 44X^6 - 136X^5 + 230X^4 - 136X^3 + 44X^2 + 8X + 1.$$

We can then compute the modified heights as before and plug them into the formula for $c_8$ from Appendix A.1 to find

$$\log|\Lambda| \geq -c_8 \log(3A/2),$$

with (as before)

$$c_8 = 1.902 \cdot 10^{20}$$

Hence we can conclude that in Case C (with $i = 2$) we have again that if $A \geq A_3 = 1.866$ then

$$A \leq A_4 = 2.8 \cdot 10^{22}.$$

Proceeding similarly to the earlier case we find with $C = 10^{40}$ that we can reduce this upper bound on $A$ to 49. Then applying the method again with $C = 10^4$ we find that $A \leq 7$.

Hence in all cases we can deduce that $A \leq 8$. It remains to check which exponents below this bound give rise to solutions, but this is trivial given the small bound on $A$. So we find the only solutions in integers $X, Y$ to the equation

$$X^4 - 2Y^4 = \pm 1$$

are given by

$$\pm(X,Y) = (1,0), (1,-1), (1,1).$$

## VII.3. The method of Bilu and Hanrot

In [**11**], Bilu and Hanrot show that for Thue equations we do not need to use the reduction technique given earlier. Instead we can use an inhomogenous linear form in two variables. We still deduce an upper bound on $A$ as before, using the theory of linear forms in logarithms, but now the reduction of the upper bound proceeds in a more efficient manner.

Let $U_I$ denote the matrix of Lemma VII.2. We let $\{i\} = \{1, \dots, s+t\} \setminus I$ and put $(u_{i,j}) = U_I^{-1}$, we shall assume that we have chosen $i$ as before so that

$$|\beta^{(i)}| = \min_{1 \leq l \leq r+1} |\beta^{(l)}|.$$

From the equation

$$U_I \begin{pmatrix} a_1 \\ \vdots \\ a_r \end{pmatrix} = \begin{pmatrix} \log \left| \frac{x-\theta^{(i_1)}y}{\mu^{(i_1)}} \right| \\ \vdots \\ \log \left| \frac{x-\theta^{(i_r)}y}{\mu^{(i_r)}} \right| \end{pmatrix}$$

we deduce, for $k = 1, \dots, r$, that

$$\begin{aligned} a_k &= \sum_{j=1}^{r} u_{k,j} \log \left| \frac{x - \theta^{(i_j)}y}{\mu^{(i_j)}} \right| \\ &= \left( \sum_{j=1}^{r} u_{k,j} \right) \log|y| + \sum_{j=1}^{r} u_{k,j} \log \left| \frac{x/y - \theta^{(i_j)}}{\mu^{(i_j)}} \right| \\ &= \delta_k \log|y| + \lambda_k + \sum_{j=1}^{r} u_{k,j} \log \left| \frac{x/y - \theta^{(i_j)}}{\theta^{(i)} - \theta^{(i_j)}} \right|, \end{aligned}$$

where

$$\begin{aligned} \delta_k &= \sum_{j=1}^{r} u_{k,j}, \\ \lambda_k &= \sum_{j=1}^{r} u_{k,j} \log \left| \frac{\theta^{(i)} - \theta^{(i_j)}}{\mu^{(i_j)}} \right|. \end{aligned}$$

Now as, if $y$ is large enough, then we have the inequality

$$\log \left| \frac{x/y - \theta^{(i_j)}}{\theta^{(i)} - \theta^{(i_j)}} \right| \le \log \frac{n+2}{n+1}$$

which means that we can estimate the absolute value of the last term in the expression for $a_k$ given above,

$$\left| \sum_{j=1}^{r} u_{k,j} \log \left| \frac{x/y - \theta^{(i_j)}}{\theta^{(i)} - \theta^{(i_j)}} \right| \right| \le \sum_{j=1}^{r} |u_{k,j}| \log \frac{n+2}{n+1} \le \frac{1}{n} \sum_{j=1}^{r} |u_{k,j}|.$$

It then follows that $A \le c_{10} \log|y| + c_{11}$, with

$$\begin{aligned} c_{10} &= \max_{1 \le k \le r} |\delta_k|, \\ c_{11} &= \max_{1 \le k \le r} \left( \frac{1}{n} \sum_{j=1}^{r} |u_{k,j}| + |\lambda_k| \right). \end{aligned}$$

Hence

$$|y|^{-1} \le c_{12} e^{-c_{13} A}$$

where $c_{12} = \exp(c_{11}/c_{10})$ and $c_{13} = c_{10}^{-1}$. This allows us to see, if $y$ is large enough, that

$$\begin{aligned}
\left|\log\left|\frac{x/y-\theta^{(i_j)}}{\theta^{(i)}-\theta^{(i_j)}}\right|\right| &\leq \left|\log\left(1+\frac{x/y-\theta^{(i)}}{\theta^{(i)}-\theta^{(i_j)}}\right)\right| \\
&\leq 2\left|\frac{x/y-\theta^{(i)}}{\theta^{(i)}-\theta^{(i_j)}}\right| \\
&\leq c_1\left|x/y-\theta^{(i)}\right| \\
&\leq c_1c_9|y|^{-n} \text{ by Lemma VII.3} \\
&\leq c_1c_9c_{12}^n e^{-nc_{13}A} = c_{14}e^{-c_{15}A}.
\end{aligned}$$

We then notice that we have the inequality

$$\begin{aligned}
|\delta_k \log|y| - a_k + \lambda_k| &\leq \left|\sum_{j=1}^{r} u_{k,j}\log\left|\frac{x/y-\theta^{(i_j)}}{\theta^{(i)}-\theta^{(i_j)}}\right|\right| \\
&\leq c_{16}e^{-c_{15}A},
\end{aligned}$$

for all $k \in \{1,\dots,r\}$, with

$$c_{16} = c_{14}\max_{1\leq k\leq r}\sum_{j=1}^{r}|u_{k,j}| = c_{14}\|U_I^{-1}\|_\infty.$$

We are now in a position to write down the inequality that Bilu and Hanrot use to reduce the bounds for Thue equations. First we define $h$ to be an index such that

$$|\delta_h| = \max_{1\leq k\leq r}|\delta_k|,$$

we then let $g$ denote any index not equal to $h$. Set $\delta = \delta_g/\delta_h$ and $\lambda = (\delta_g\lambda_h - \delta_h\lambda_g)/\delta_h$ then

$$\begin{aligned}
|a_g - \delta a_h + \lambda| &= |(\delta_g\log|y| - a_g + \lambda_g) - \delta\,(\delta_h\log|y| - a_h + \lambda_h)| \\
&\leq |\delta_g\log|y| - a_g + \lambda_g| + |\delta|\;|\delta_h\log|y| - a_h + \lambda_h| \\
&\leq (1+|\delta|)c_{16}e^{-c_{15}A} \\
&\leq 2c_{16}e^{-c_{15}A}.
\end{aligned}$$

We can use this later inequality to reduce the bound on $A$ using LLL–reduction of two dimensional lattices. However, it is more efficient to use continued fractions, just as we did in the homogeneous case in Exercise V, 6. Hence no matter what the degree of the equation we are solving we can use the continued fraction algorithm to reduce the bounds. This dispenses with the need to use large dimensional lattices.

We can also use the last inequality to search for the small solutions to the equation when we cannot reduce the bounds on $A$ any further. The 'naive method', given an upper bound $A_6$ on $A$, would loop through all possible values of the exponents. This would mean we would need to check

$$(2A_6 + 1)^r$$

possibilities. As mentioned earlier this can be reduced by using a sieving method on the exponents. However, rewriting the above inequalities to deduce right hand sides depending on $|y|$ rather than $A$, we can deduce that if $|y|$ is large enough we will obtain

$$|a_g - \frac{1}{\delta_h}(\delta_g a_h - \delta_g \lambda_h + \delta_h \lambda_g)| < 1/2.$$

So we loop through all possible values for $a_h$, and then the possible values of $a_g$ for all $g \neq h$ are determined explicitly. Hence we need only check $2A_6 + 1$ exponent vectors. This means that the final search depends only on $A_6$ and not on $r$ and hence the degree of the equation.

However, this does not get around the main computational bottleneck of the whole procedure, which is the computation of the number field data. But assuming we can compute the fundamental units and a complete set of coset representatives for the $\mu$ then the whole procedure for solving a Thue equation is very fast and does not essentially depend on the degree of the equation. In [**12**] this method is used to solve a Thue equation of degree 2505.

## VII.4. Integral points on elliptic curves (I)

We are now able to describe our first general method for finding integral points on an elliptic curve. The method is the classical one which can be found in Mordell's book [**138**]. We assume that our elliptic curve is given by an equation of the form

$$Y^2 = F(X) \tag{VII.5}$$

where $F(X) \in \mathbb{Z}[X]$ is a monic, cubic polynomial with non-zero discriminant. We wish to determine all $(X, Y) \in \mathbb{Z}^2$ which satisfy (VII.5). Let $L$ denote the algebra $\mathbb{Q}[X]/(F(X))$; this is nothing but one of the following three sums of fields:

1. $F(X)$ has three integer roots:
$$L = \mathbb{Q} + \mathbb{Q} + \mathbb{Q}.$$
2. $F(X)$ has one integer root and one quadratic root $\theta$:
$$L = \mathbb{Q} + \mathbb{Q}(\theta).$$
3. $F(X)$ is irreducible with $F(\theta) = 0$ for some $\theta \in \mathbb{C}$:
$$L = \mathbb{Q}(\theta).$$

Write

$$L = \sum_{i=1}^{r} L_i$$

where $L_i = \mathbb{Q}(\theta_i)$ is a number field, and $\theta_i$ is a corresponding root of the cubic $F(X)$. Suppose that $(X, Y)$ is an integral point on our elliptic curve, then $X - \theta_i \in L_i$ is an algebraic integer in $L_i$. If $\mathfrak{p}$ is a prime ideal divisor of the ideal $(X - \theta_i)$ to an odd degree, then by the unique factorization of ideals we have, from the equation

$$(Y)^2 = (X - \theta_1)(X - \theta_2)(X - \theta_3),$$

that $\mathfrak{p}$ also divides $(X - \theta_j)$ for some $j \neq i$. Hence $\mathfrak{p}$ must divide the discriminant of $F(X)$ as $\mathfrak{p}$ divides $(X - \theta_j) - (X - \theta_i) = (\theta_i - \theta_j)$.

If we let $S_i$ denote the set of prime ideals of $L_i$ which divide the discriminant of $F(X)$ then it is clear that we have that

$$X - \theta_i = \alpha_i \beta_i^2$$

where

1. $\alpha_i \in L_i(S_i, 2)$.
2. $\beta_i \in L_i$
3. $\prod_{i=1}^{r} N_{L_i/\mathbb{Q}}(\alpha_i) =$ a rational square.

Remember, from Chapter IV, that $L_i(S_i, 2)$ denotes the set of elements of $L_i^*/L_i^{*2}$ which will give, upon adding the square root of an element of $L_i(S_i, 2)$, an extension unramified away from $S_i$. We take representatives for elements of $L_i(S_i, 2)$ to be integral elements with no square divisors in $\mathcal{O}_K/\mathcal{O}_K^*$. So, for a finite number of possibilities for the $\alpha_i$, we are reduced to finding all the possible values of $\beta_i$.

LEMMA VII.6. *In such a situation there exist three quadratic forms, $Q_1$, $Q_2$, $Q_3$, with integral coefficients in three variables $x, y, z$ such that solving our integral point problem means finding integer solutions, $(x, y, z)$, to the simultaneous system*

$$Q_1(x, y, z) = 0 \ , \ Q_2(x, y, z) = 1 \ , \ Q_3(x, y, z) = X$$

PROOF. Exercise. □

Now the equation $Q_1(x, y, z) = 0$ is a curve of genus zero in $\mathbb{P}^2$ so by Theorem III.1 it has a rational solution if and only if it has a solution in every $\mathbb{Q}_p$. The existence of $\mathbb{Q}_p$ solutions is easy to determine, as has already been discussed in Chapter IV. In Chapter IV we also discussed how given a rational solution we can parameterize all solutions to $Q_1(x, y, z) = 0$ in terms of three quadratic forms in two variables

$$gx = q_1(m, n) \ , \ gy = q_2(m, n) \ , \ gz = q_3(m, n),$$

where $m$ and $n$ are coprime integer variables and $g$ is an integer constant which comes from some finite set and the $q_i$ have integral coefficients.

Then substituting the $q_i(m,n)$ into $Q_2(x,y,z) = 1$ we obtain a Thue equation of degree four:

$$G(m,n) = g^2,$$

(strictly it is a Thue equation only when $g \neq 0$. When $g = 0$ it is easy to solve). Hence by the methods of this chapter we can find all the solutions, $(m,n)$. We can then find all values of $x, y, z$ using the quadratic forms $q_i(m,n)$ and finally we can find the possible integral points $(X,Y)$ from the equation $X = Q_3(x,y,z)$.

Before we continue with an example one should note that the above procedure for finding integral points is closely related to the procedure known as 2-descent for finding the structure of the rational points on an elliptic curve. We shall return to 2-descent in Chapter XII.

## VII.5. $Y^2 = X^3 - 6X - 14$

Suppose we wish to find all integral points on the elliptic curve

$$Y^2 = F(X) = X^3 - 6X - 14.$$

If we set $K = \mathbb{Q}(\theta)$ where $\theta^3 - 6\theta - 14 = 0$ then we find that $K$ has class number one, one fundamental unit which we can take to be $\eta = (5-8\theta+2\theta^2)/3$ and an integral basis given by $\omega_1 = 1, \omega_2 = \theta$ and $\omega_3 = (1 - \theta + \theta^2)/3$. The discriminant of $F(X)$ is $-2^2 \cdot 3^3 \cdot 41$ which means that we need to find generators of the prime ideals lying above 2, 3 and 41. We find that these ideals split as

$$(2) = \mathfrak{p}_2^3\ ,\ (3) = \mathfrak{p}_3\mathfrak{p}_3'^{\,2}\ ,\ (41) = \mathfrak{p}_{41}\mathfrak{p}_{41}'^{\,2}.$$

As generators we can take (with an obvious correspondence)

$$\begin{array}{ll} & \pi_2 = (8+\theta-\theta^2)/3, \\ \pi_3 = (5+\theta-\theta^2)/3, & \pi_3' = (1+2\theta+\theta^2)/3, \\ \pi_{41} = (1+5\theta-2\theta^2)/3, & \pi_{41}' = 5+4\theta+\theta^2. \end{array}$$

Clearly we then have that

$$K(S,2) \cong \langle -1 \rangle \times \langle \eta \rangle \times \langle \pi_2 \rangle \times \langle \pi_3 \rangle \times \langle \pi_3' \rangle \times \langle \pi_{41} \rangle \times \langle \pi_{41}' \rangle,$$

which is a group of order $2^7$. But we are only interested in those elements whose norms are squares in $\mathbb{Q}$. A simple calculation then reveals that we are therefore only interested in the subgroup of order 8 given by

$$H \cong \langle -\eta \rangle \times \langle -\pi_3\pi_3' \rangle \times \langle -\pi_{41}\pi_{41}' \rangle.$$

For each $\alpha \in H$ we then have to determine whether there are integers $x, y, z$ such that

$$X - \theta = \alpha(x\omega_1 + y\omega_2 + z\omega_3)^2.$$

Equating the coefficients of $\theta$ in both side of this inequality gives us

$$0 = Q_1(x,y,z)\ ,\ -1 = Q_2(x,y,z)\ ,\ X = Q_3(x,y,z).$$

We take the first quadratic form first,

$$Q_1(x, y, z) = 0.$$

We have eight possibilities for $Q_1$ depending on our choice of $\alpha$. These are exactly the eight quadratic forms which you were asked to study in Chapter IV, Exercises 3 and 4. Clearly a necessary condition is that $Q_1(x, y, z) = 0$ must be locally soluble everywhere. You should have found that three out of the eight quadratic forms were not locally soluble everywhere. Eliminating these means that we can restrict ourselves to $\alpha$ coming from the set

$$H' = \{1, -\pi_3\pi_3', \eta\pi_3\pi_3', \eta\pi_{41}\pi_{41}', -\eta\pi_3\pi_3'\pi_{41}\pi_{41}'\}.$$

For the elements in $H'$ we find a solution to $Q_1(x, y, z)$ and then write $x, y, z$ as quadratic forms in two integer coprime integer variables. This gives rise to a Thue equation of degree four which we then need to solve using the techniques of this chapter. We now take each possible value of $\alpha$ in turn:

**i)** $\alpha = 1$. We find

$$Q_1(x, y, z) = (2xz + 3y^2 - 2yz + 3z^2)/3 = 0$$

and then

$$gx = -3p^2 + 2pq - 3q^2, \;\; gy = 2pq, \;\; gz = 2q^2.$$

Clearly we then can choose $g = 1, 2, 3$ or $6$ by the method of Chapter IV. We then substitute these quadratic forms into the equation

$$Q_2(x, y, z) = (-2xz + 6xy + 14yz)/3 = -1$$

to find the Thue equation:

$$4q(q^3 + pq^2 + 3p^2q - 3p^3) = -g^2,$$

which clearly has no integer solutions if $g = 1$ or $3$. For the other values of $g$ this only has a solution when $g = 6$ in which case we have $(p, q) = \pm(2, 1)$. But then $(x, y, z) = (-11/6, 2/3, 1/3)$ which is not integral.

**ii)** $\alpha = -\pi_3\pi_3'$. Here we find

$$Q_1(x, y, z) = 2xy + y^2 + 4yz + z^2 = 0$$

and

$$gx = -q^2 - p^2 - 4pq, \;\; gy = 2p^2, \;\; gz = 2pq.$$

We can then choose $g = 1$ or $2$, in the Thue equation

$$q^4 - 6p^2q^2 + 24p^3q + 21p^4 = -g^2,$$

which we then find has no solutions.

**iii)** $\alpha = \eta\pi_3\pi_3'$. Here we find

$$Q_1(x, y, z) = 2x^2 - 6yx + 4zx + y^2 + 4zy - 5z^2 = 0$$

and

$$gx = p^2 + 4pq - 5q^2,\ gy = p^2 + 6pq - 5q^2,\ gz = p^2 + 4pq - 3q^2.$$

We can then choose $g = 1$ or 2, in the Thue equation

$$2(p^4 + 6p^2q^2 - 12pq^3 + 3q^4) = -g^2,$$

which has the solutions $(p, q, g) = (1, 1, 2), (-1, -1, 2)$. Then we have that $(x, y, z)$ must be equal to $(0, 1, 1)$, and this gives the integral point $(X, Y) = (5, \pm 9)$.

**iv)** $\alpha = \eta\pi_{41}\pi'_{41}$. We now have

$$Q_1(x, y, z) = (15x^2 + 18xy - 58xz - 93y^2 + 244yz - 138c^2)/3 = 0$$

and

$$gx = 93p^2 - 244pq + 138q^2,\ gy = 6163p^2 - 10270pq + 4278q^2,$$
$$gz = 3441p^2 - 5748pq + 2400q^2.$$

Taking the Hermite normal form of the resulting matrix we see that we can restrict $g$ to come from the set $\{1, 2, 3, 6, 41, 82, 123, 246\}$. For each of these we have to solve the Thue equation

$$82(6708639p^4 - 22158184p^3q + 27445440p^2q^2 - 15108756q^3p + 3119058q^4) = -g^2.$$

We can therefore restrict $g$ even more; it must either be 82 or 246. We find that there are no solutions to these two Thue equations.

**v)** $\alpha = -\eta\pi_3\pi'_3\pi_{41}\pi'_{41}$. In this final case we have

$$Q_1(x, y, z) = 8x^2 - 56xy + 62xz + 57y^2 - 94yz + 33z^2 = 0$$

which leads us to

$$gx = 57p^2 - 94pq + 33q^2,\ gy = 392p^2 - 458pq + 132q^2,$$
$$gz = 342p^2 - 400pq + 116q^2.$$

The Hermite normal form of the resulting matrix tells us that $g$ must lie in the set $\{1, 2, 41, 82\}$. For each of these four values we need to solve the Thue equation

$$164(5636p^4 - 14809p^3q + 14211p^2q^2 - 5941pq^3 + 917q^4) = -g^2.$$

Hence $g$ must be equal to 82, and we find that there are no solutions to this equation.

In summary the only integral points on the elliptic curve

$$Y^2 = X^3 - 6X - 14$$

are given by $(X, Y) = (5, \pm 9)$.

## VII.6. Exercises

1). Verify all the numerical calculations in Section VII.2.

2). Prove Lemma VII.6.

3). Work out all the details for solving the Thue equations which arose when we solved $Y^2 = X^3 - 6X - 14$.

4). Find all the integral points on the curves
(i) $Y^2 = X^3 + 3$.
(ii) $Y^2 = X^3 - 52$.
(iii) $Y^2 = X^3 + X + 1$.
(iv) $Y^2 = X^4 - 3X + 1$.

5). If our reduced bound on $A$ is not small enough then it may be hard to perform a naive search to find all the solutions to a Thue equation. Give a method using continued fractions to overcome this problem.

CHAPTER VIII

# Thue–Mahler equations

## VIII.1. Thue–Mahler equations

The obvious way to generalize the Thue equation is to consider a Thue–Mahler equation. In a Thue–Mahler equation we replace the fixed integer $m$ on the right hand side by a product of a fixed integer and an element from a multiplicative, finitely generated sub-coid of $\mathbb{Z}$. (A 'coid' is a set with a binary operation which is associative, is **CO**mmutative and has an **ID**entity, a terminology I am grateful to D. Bernstein for pointing out to me. For those who prefer more standard notation we mean nothing more than a commutative monoid.) In practice we can take the generators of the coid to be prime numbers, hence our Thue–Mahler equation is

$$F(X,Y) = Cp_1^{z_1} \cdots p_t^{z_t}.$$

We now wish to determine not only $X$ and $Y$ but also the $z_i$. It is clear that the only way we can obtain finitely many solutions is to consider only those solutions with $(X,Y) = 1$. We can assume, without loss of generality, that $(m, p_i) = 1$ and that $F(X,Y)$ is a monic, irreducible form of degree greater than 2.

That a Thue–Mahler equation has only finitely many solutions was first proved by Mahler [**124**] in 1933. However, Mahler's proof was non-effective. It was Coates [**30**] who in 1969, by generalizing Baker's work to the $p$-adic case, was able to give an effective proof of the finiteness of the number of solutions. In 1992 Tzanakis and de Weger [**204**] gave a completely general practical method for solving such equations by combining Baker's theory (and its $p$-adic analogues) with LLL and a sieving technique.

We give a rough sketch of the method of Tzanakis and de Weger. Those wishing to implement the method should consult their original paper as it contains many optimizations and discussions that aid the overall process.

We proceed as before, setting $\theta$ to be a root of $F(X,1)$. We first wish to consider the prime ideal factorization of $(X - \theta Y)$ for a solution pair $(X,Y)$. This clearly could be divisible (to some arbitrary power) by every prime ideal, $\mathfrak{p}$, of $K = \mathbb{Q}(\theta)$ lying above one of the $p_i$. The first simplification that Tzanakis and de Weger make is to show that one does not have complete freedom here. They eliminate many possibilities by appealing to their *prime ideal removing lemma.*

## VIII.2. The prime ideal removing lemma

We fix a rational prime $p$. For $i = 1, \dots, m$ we let $\mathfrak{p}_i$ denote the prime ideals of $K$ lying above $p$, with ramification index $e_i$ and residue degree $f_i$. We let $g_i(X)$ denote the $p$-adic factor of $F(X, 1)$ corresponding to the prime ideal $\mathfrak{p}_i$ and let the $p$-adic roots of $g_i(X)$ be labelled $\theta_i^{(j)}$, for $j = 1, \dots, n_i = e_i f_i$. We assume throughout that $X$ and $Y$ are coprime integers.

LEMMA VIII.1. *For every pair $i, j \in \{1, \dots, m\}$ with $i \neq j$ there is at most one prime ideal $\mathfrak{p} \in \{\mathfrak{p}_i, \mathfrak{p}_j\}$ such that*

$$\mathrm{ord}_{\mathfrak{p}}(X - \theta Y) > \max\{e_i, e_j\} \mathrm{ord}_p(\theta_i^{(k)} - \theta_j^{(l)}). \tag{VIII.1}$$

*The indices $k, l$ are arbitrarily chosen from the sets $\{1, \dots, n_i\}$ and $\{1, \dots, n_j\}$ respectively.*

PROOF. All we need to show is that if, for some integral ideal $\mathfrak{a}$,

$$(X - \theta Y) = \mathfrak{p}_i^{v_i} \mathfrak{p}_j^{v_j} \mathfrak{a}$$

then

$$\min\{v_i, v_j\} \leq \max\{e_i, e_j\} \mathrm{ord}_p(\theta_i^{(k)} - \theta_j^{(l)}).$$

It is clear that

$$\frac{\min\{v_i, v_j\}}{\max\{e_i, e_j\}} \leq v_i / e_i \leq \frac{1}{e_i} \mathrm{ord}_{\mathfrak{p}_i}(X - Y\theta) = \mathrm{ord}_p(X - Y\theta_i^{(k)}),$$

with a similar inequality holding for $\mathrm{ord}_p(X - Y\theta_j^{(l)})$. From these two inequalities we deduce that

$$\begin{aligned} \mathrm{ord}_p\left(Y\left(\theta_i^{(k)} - \theta_j^{(l)}\right)\right) &\geq \min\left\{\mathrm{ord}_p(X - Y\theta_i^{(k)}), \mathrm{ord}_p(X - Y\theta_j^{(l)})\right\} \\ &\geq \frac{\min\{v_i, v_j\}}{\max\{e_i, e_j\}}. \end{aligned}$$

Hence if $\mathrm{ord}_p(Y) = 0$ then we are done, so suppose that $p$ divides $Y$. Now as $X$ and $Y$ are coprime it follows that $p$ does not divide $X$. Hence for any $\mathfrak{p}$ dividing $p$ we have

$$\mathrm{ord}_{\mathfrak{p}}(X - Y\theta) = 0$$

and so $v_i = v_j = 0$ and the result follows. □

LEMMA VIII.2. *Suppose equation (VIII.1) holds for $\mathfrak{p} = \mathfrak{p}_i$ and either $e_i > 1$ or $f_i > 1$ then*

$$\mathrm{ord}_{\mathfrak{p}_i}(X - \theta Y) \leq e_i \mathrm{ord}_p(\theta_i^{(k)} - \theta_i^{(l)}) \tag{VIII.2}$$

*Here the indices $k, l$ are distinct indices chosen from the set $\{1, \dots, n_i\}$.*

PROOF. Write

$$(X - Y\theta) = \mathfrak{p}_i^{v_i} \mathfrak{a}$$

for some integral ideal $\mathfrak{a}$. We have that $n_i = e_i f_i > 1$ by assumption and so there are certainly two distinct indices $k, l$ in the set $\{1, \ldots, n_i\}$. We can apply the argument used in the previous lemma to the equalities

$$\operatorname{ord}_p(X - Y\theta_i^{(k)}) = \frac{v}{e_i}\ , \ \operatorname{ord}_p(X - Y\theta_i^{(l)}) = \frac{v}{e_i}$$

to deduce the result. □

COROLLARY VIII.3. *There is at most one prime ideal* $\mathfrak{p}_i$ *dividing* $p$ *with*

$$\operatorname{ord}_{\mathfrak{p}_i}(X - \theta Y) > \max_{1 \le j < k \le m} \left( \max\{e_j, e_k\} \operatorname{ord}_p(\theta_j^{(h)} - \theta_k^{(l)}) \right), \tag{VIII.3}$$

*where* $h \in \{1, \ldots, n_j\}$ *and* $l \in \{1, \ldots, n_k\}$ *are arbitrary. Moreover if (VIII.3) holds with* $e_i > 1$ *or* $f_i > 1$ *then* $\mathfrak{p}_i$ *also satisfies (VIII.2).*

PROOF. This is just the two lemmas stated in other language. □

COROLLARY VIII.4. *There is at most one prime ideal* $\mathfrak{p}_i$ *dividing* $p$ *with*

$$\operatorname{ord}_{\mathfrak{p}_i}(X - \theta Y) > \frac{\max\{e_1, \ldots, e_m\}}{2} \operatorname{ord}_p D(\theta),$$

*such a* $\mathfrak{p}_i$ *must also satisfy* $e_i = f_i = 1$. *Here* $D(\theta)$ *is the discriminant of* $\theta$ *as an algebraic integer over* $\mathbb{Q}$.

PROOF. Exercise. □

The last corollary is also applied in the number field sieve factoring method. It is the reason that only degree one prime ideals need be considered in the algebraic factor base. For more information on factoring and the number field sieve you should consult [**32**] and [**116**].

What the prime ideal removing lemma has told us is that if a prime ideal divides $X - \theta Y$ to a positive power then it is a degree one prime ideal or it divides the discriminant of $\theta$. In either case at most one prime ideal lying above a given rational prime $p$ can divide $X - \theta Y$ to some arbitrary power and such a prime ideal must be of degree one and unramified. Hence this reduces considerably the numbers of prime ideals which can divide the principle ideal $(X - \theta Y)$ to an unbounded power.

## VIII.3. The method

To recap we are trying to solve the equation

$$F(X, Y) = C p_1^{z_1} \cdots p_t^{z_t}.$$

Using unique factorization of ideals and the prime ideal removing lemma we have a finite number of cases of the form (where as before $F(\theta, 1) = 0$)

$$\beta = X - \theta Y = \mu \epsilon \pi_1^{s_1} \cdots \pi_t^{s_t}$$

where

- $\mu \in \mathcal{O}_K$ is from a finite set which can be effectively determined,

- $\epsilon$ a unit of $K$,
- $(\pi_i) = \mathfrak{p}_i^{h_i}$ with $\mathfrak{p}_i$ an unramified degree one prime ideal lying above $p_i$.

The number $h_i$ is the smallest positive integer such that $\mathfrak{p}_i^{h_i}$ is principal, so it is clear that $1 \leq h_i \leq h_K$. Note if no such $\mathfrak{p}_i$ lies above $p_i$ then we can bound the corresponding $z_i$ by the prime ideal removing lemma and hence absorb this factor into $m$. We shall write $\epsilon$ (as in the case of Thue equations) as

$$\epsilon = \eta_1^{a_1} \cdots \eta_r^{a_r}$$

where $\eta_1, \ldots, \eta_r$ are a set of fundamental units for $K$, by absorbing the units of finite order into $\mu$.

We set $A = \max(|a_i|)$, $S = \max(|s_i|)$ and $H = \max(A, S)$ hence it is enough to find a bound on $H$ to solve our equation. In practice we will require a 'small' bound on $H$ and a means to evaluate all values of $a_i$ and $s_i$ with $|a_i|, |s_i| \leq H$ in an efficient manner. Writing

$$\tau_1 = \prod_{l=1}^{r} \left( \frac{\eta_l^{(i)}}{\eta_l^{(j)}} \right)^{a_l} \prod_{l=1}^{t} \left( \frac{\pi_l^{(i)}}{\pi^{(j)}} \right)^{s_l}, \quad \tau_2 = \prod_{l=1}^{r} \left( \frac{\eta_l^{(k)}}{\eta_l^{(j)}} \right)^{a_l} \prod_{l=1}^{t} \left( \frac{\pi_l^{(k)}}{\pi^{(j)}} \right)^{s_l}$$

and defining $\alpha_1, \alpha_2$ as in the case of Thue equations (see equation (VII.3)) we have the identity

$$\alpha_1 \tau_1 + \alpha_2 \tau_2 + 1 = 0.$$

Again notice how the $\alpha_i$ are fixed and the $\tau_i$ range over two finitely generated subgroups of the algebraic numbers.

Just as in the case of Thue equations we define

$$\begin{aligned} c_1 &= 2 \left( \min_{l \neq m} |\theta^{(m)} - \theta^{(l)}| \right)^{-1} \\ c_2 &= \max_{l_1 \neq l_2 \neq l_3 \neq l_1} \left| \frac{(\theta^{(l_2)} - \theta^{(l_3)})}{(\theta^{(l_3)} - \theta^{(l_1)})} \right|. \end{aligned}$$

Then we have $|\alpha_1 \tau_1| \leq c_1 c_2 |\beta^{(i)}| = c_3 |\beta^{(i)}|$, where as before we choose the conjugate $i \in \{1, \ldots, n\}$ so that

$$|\beta^{(i)}| = \min_{1 \leq l \leq n} |\beta^{(l)}|$$

and again we choose $j, k$ to be two distinct arbitrary indices not equal to $i$. However, there are two important differences with the Thue equation approach. For Thue equations we saw that we could restrict the index $i$ to correspond to a real embedding; this is now no longer possible. For Thue equations we also split the discussion into three subcases, the first two of which provided trivial bounds on the exponents. Here we will again obtain three subcases but this time we have no trivial bounds.

Also, as in the case of Thue equations, we use Lemma VII.2 to show that there is an index $h$ and a constant $c_4$ such that

$$|\log |\epsilon^{(h)}|| \geq c_4 A.$$

We choose $c_5$ as we did for Thue equations and split the discussion into various cases. However, first we shall use non-archimedian information to show that $S$ can always be bounded in terms of $H$.

LEMMA VIII.5. *There are constants $c_6$ and $c_7$ such that*

$$S \leq c_6(\log H + c_7).$$

PROOF. First let $p_l$ denote any one of our unbounded primes, with the degree one prime ideal $\mathfrak{p}_l$ lying above it. We wish to show that there are constants $c_8(p_l)$ and $c_9(p_l)$, which depend on our choice of rational prime, so that

$$s_l \leq c_8(p_l)(\log H + c_9(p_l)).$$

It is clear that for any choice of indices $i_l, j_l, k_l$ we must have

$$\mathrm{ord}_{p_l}\left(\frac{\eta_l^{(k_l)}}{\eta_l^{(j_l)}}\right) = \mathrm{ord}_{p_l}\left(\frac{\eta_l^{(i_l)}}{\eta_l^{(j_l)}}\right) = 0.$$

We choose the three indices in the expressions for $\tau_1$ and $\tau_2$ above so that

$$\mathrm{ord}_{p_l}(\alpha_2) \geq 0\ ,\ \mathrm{ord}_{p_l}\left(\frac{\pi_l^{(k_l)}}{\pi_l^{(j_l)}}\right) = 0 \text{ and } \mathrm{ord}_{p_l}\left(\frac{\pi_l^{(i_l)}}{\pi_l^{(j_l)}}\right) = h_l.$$

We leave it as an exercise to show that this can always be done. With this choice of indices we see that

$$\mathrm{ord}_{p_l}(\alpha_1\tau_1) = \mathrm{ord}_{p_l}(\alpha_1) + h_l s_l \text{ and } \mathrm{ord}_{p_l}(\alpha_2\tau_2) = \mathrm{ord}_{p_l}(\alpha_2)$$

Now if $\mathrm{ord}_{p_l}(\alpha_2) > 0$ then

$$s_l = \frac{-1}{h_l}\mathrm{ord}_{p_l}(\alpha_1).$$

So we can assume that $\mathrm{ord}_{p_l}(\alpha_2) = 0$, but then we can apply Yu's theorem, see Appendix A.2, to find constants $c_{10}(p_l)$ and $c_{11}(p_l)$ so that

$$\mathrm{ord}_p(-\alpha_2\tau_2 - 1) \leq c_{10}(p_l)(\log H + c_{11}(p_l)).$$

Hence combining this with the equality

$$h_l s_l = \mathrm{ord}_p(-\alpha_2\tau_2 - 1) - \mathrm{ord}_{p_l}(\alpha_1)$$

we achieve our inequality of the form

$$s_l \leq c_8(p_l)(\log H + c_9(p_l)).$$

The result required by the lemma then follows on taking the maximum of the constants in all cases. □

Note that the above lemma and Lemma II.9 show that, if $s_l > 1$, then there is a linear form in $p$-adic logarithms, $\Lambda_p$, such that

$$\mathrm{ord}_p(\Lambda_p) = h_l s_l + \mathrm{ord}_{p_l}(\alpha_1)$$

i.e.

$$|\Lambda_p|_p \leq p^{-h_l S - \mathrm{ord}_{p_l}(\alpha_1)} = c_{12}e^{-c_{13}S}.$$

It is this last inequality which we shall use to reduce the bound on $S$ later using the reduction theory of $p$-adic linear forms considered in Section VI.4.

If $H = S$ we are then able to bound all the variables using the above lemma. We now turn to trying to bound $H$ in the case $H = A$. To do this we must use archimedian information. From the inequality $|\alpha_1\tau_1| \le c_3|\beta^{(i)}|$, we see that if $|\beta^{(i)}|$ is small then $\Lambda = \log(-\alpha_2\tau_2)$ will also be very small. Now, as promised, we split the discussion into our three cases which arose when we considered Thue equations.

**Case A: $|\beta^{(i)}| > e^{-c_5 A}$ and $|\epsilon^{(h)}| \ge e^{c_4 A}$.** It is easy to see that in this case we have the inequality

$$|\beta^{(h)}| < C p_1^{z_1} \cdots p_t^{z_t} e^{(n-1)c_5 A}.$$

Remembering that $\pi_l$ for $l = 1, \dots, t$ is the generator of the $h_l$th power of a degree one prime ideal we see that $z_l = h_l s_l + t_l$, where $t_l$ is given by

$$t_l = \mathrm{ord}_p\left(N_{K/\mathbb{Q}}(\mu)/C\right).$$

From this we can deduce the inequality

$$\begin{aligned} e^{c_4 A} &\le |\epsilon^{(h)}| = \frac{|\beta^{(h)}|}{|\mu^{(h)}| \prod_{l=1}^{t} |\pi_l^{(h)}|^{s_l}}, \\ &\le \frac{C}{|\mu^{(h)}|} e^{(n-1)c_5 A} \prod_{l=1}^{t} \left( p_l^{t_l} \left| \frac{p_l^{h_l}}{\pi_l^{(h)}} \right|^{s_l} \right). \end{aligned}$$

We then define the constants

$$\begin{aligned} c_{14} &= \max_{1 \le l \le t} \left( \max_{1 \le g \le n} \log \left| \frac{p_l^{h_l}}{\pi_l^{(g)}} \right| \right), \\ c_{15} &= \max_{1 \le l \le n} \log \left( \frac{C}{|\mu^{(l)}|} \prod_{l=1}^{t} p_l^{t_l} \right). \end{aligned}$$

We then deduce the inequality

$$c_4 A \le c_{15} + (n-1)c_5 A + c_{14} S$$

and finally the inequality

$$A = H \le \frac{c_{15} + c_{14} S}{c_4 - (n-1)c_5}.$$

From this inequality we can deduce an upper bound on $H$ using the inequality for $S$ from Lemma VIII.5 and Appendix B.1:

$$H \le \frac{2}{c_4 - (n-1)c_5} \left( c_{15} + c_{14} c_6 c_7 + c_{14} c_6 \log \left( \frac{c_{14} c_6}{c_4 - (n-1)c_5} \right) \right).$$

**Case B:** $|\beta^{(i)}| > e^{-c_5 A}$ **and** $|\epsilon^{(h)}| \leq e^{-c_4 A}$**.** We immediately deduce the inequality

$$e^{-c_4 A} \geq |\epsilon^{(h)}| > \frac{|\beta^{(i)}|}{|\mu^{(h)} \prod_{l=1}^{t} \pi_l^{(h)s_l}|}.$$

Upon setting

$$\begin{aligned} c_{16} &= \max_{1\leq l\leq n} \log|\mu^{(l)}|, \\ c_{17} &= \max_{1\leq l\leq n} \log|\prod_{v=1}^{s} \pi_v^{(l)}|, \end{aligned}$$

we deduce the inequality

$$-c_4 A > -c_5 A - c_{16} - c_{17} S$$

and finally the inequality

$$A \leq \frac{c_{16} + c_{17} S}{c_4 - c_5}.$$

Again using Lemma VIII.5 to bound $S$ in terms of $\log H$ we are able to find an upper bound on $H$ using Appendix B.1.

**Case C:** $|\beta^{(i)}| \leq e^{-c_5 A}$**.** As in the case of the Thue equation we deduce the inequality

$$|\alpha_1 \tau_1| \leq c_3 e^{-c_5 A}.$$

Now setting

$$\Lambda = \log(-\alpha_2 \tau_2)$$

we see that for sufficiently large $A$ we can bound the linear form in logarithms, $\Lambda$, with the inequality

$$|\Lambda| \leq 2c_3 e^{-c_5 A}.$$

Then using Appendix A.1 we determine a constant $c_{18}$ so that

$$-c_{18} \log((r+t+1)A/2) \leq \log|\Lambda| \leq \log(2c_3) - c_5 A.$$

This allows us to find an upper bound on $H$, using again Appendix B.1,

$$H \leq \frac{2}{c_4 - c_5}\left(c_{16} + c_{17}c_6c_7 + c_{17}c_6 \log\left(\frac{c_{17}c_6}{c_4 - c_5}\right)\right).$$

Note that in all three cases the upper bound on $H$ will be very large because they all depend on the lower bounds on $p$-adic logarithms arising from applying Yu's theorem in Lemma VIII.5. The strategy which we adopt is to reduce the bound on $S$ using our earlier discussion of reducing bounds on linear forms in $p$-adic numbers. This will then give smaller bounds in the three cases $H = S$, Case A and Case B. However, for Case C this gives no help so we need to reduce the bound in Case C using the standard reduction theory for complex linear forms.

We can then feed these reduced bounds back into the reduction process for $S$ to find a new (hopefully) smaller upper bound. We keep repeating this process of $p$-adic reduction followed by complex reduction until our upper bound on $H$ can be reduced no more.

We now have (hopefully) a small upper bound on $H$ which allows us to enumerate all the possible exponents and check whether they give a solution to our equation. Alas for Thue–Mahler equations it is not known whether we can generalize the method of Bilu and Hanrot. Hence for large degree equations we require the reduction of large matrices and a final search region which grows exponentially with the degree.

Before we present an example two things should be pointed out. Firstly the above method for solving Thue–Mahler equations requires that we only work in the field $K = \mathbb{Q}(\theta)$, so field arithmetic is rather cheap. Secondly, we have treated the archimedian and non-archimedian primes in a rather unequal way. If one directly generalizes the above method for general two term $S$-unit equations then one would soon find that this imbalance between the treatments of archimedian and non-archimedian primes would cause a problem as we no longer have a prime ideal removing lemma to remove all ideals whose degree is not equal to one. In the next chapter we do give such a method where we treat all primes equally; however, such egalitarianism is at the expense of having to use larger field extensions.

## VIII.4. $X^3 - X^2Y + XY^2 + Y^3 = \pm 11^s$

We end the discussion on Thue–Mahler equations with an example originally considered by Agrawal, Coates, Hunt and van der Poorten [**2**], in the context of finding all elliptic curves of conductor 11. They showed that the problem of determining all such elliptic curves was equivalent to finding all solutions to the following Thue–Mahler equation:

$$X^3 - X^2Y + XY^2 + Y^3 = \pm 11^s.$$

They used a rather different method from the general one considered here to solve the problem. We shall use the method we have described above.

The field $K = \mathbb{Q}(\theta)$ where $\theta^3 - \theta^2 + \theta + 1$ has discriminant $-2^2 \cdots 11$ and so has one fundamental unit which we can take to be $\theta$. The prime ideal (11) decomposes as a product of two degree one prime ideals (one of which has ramification index two). The two ideals are principal and we can take

$$(11) = \mathfrak{p}_1\mathfrak{p}_2^2\ ,\ \mathfrak{p}_1 = (\pi_1)\ ,\ \mathfrak{p}_2 = (\pi_2)$$

where

$$\pi_1 = 1 - 2\theta\ ,\ \pi_2 = 2 + \theta^2.$$

As $D(\theta) = -2^2 11$, we have $\mathrm{ord}_{11} D(\theta) = 1$ and so the prime ideal removing lemma tells us that we must have either

$$X - \theta Y = \pm\theta^{a_1}\pi_1^{s_1},$$

or

$$X - \theta Y = \pm\pi_2\theta^{a_1}\pi_1^{s_1}.$$

We shall only consider the first case here, leaving the second as an exercise. For use later on we note the following minimal polynomials and heights of various algebraic numbers.

The minimal polynomial of $\theta^{(i)}/\theta^{(j)}$ is

$$g(X) = X^6 + 4X^5 + 11X^4 + 12X^3 + 11X^2 + 4X + 1,$$

which we can take as the defining polynomial of the Galois closure, $K^{Gal}$, of $K$. From this we can determine that $h(\theta^{(i)}/\theta^{(j)}) = 0.304688$ and that the prime (11) decomposes in the Galois closure as a product of three ramified prime ideals. The minimal polynomial of $\pi_1^{(i)}/\pi_1^{(j)}$ is

$$121X^6 + 330X^5 + 599X^4 + 716X^3 + 599X^2 + 330X + 121,$$

which leads us to deduce that $h(\pi_1^{(i)}/\pi_1^{(j)}) = 0.830993$. The minimal polynomial of $(\theta^{(i)} - \theta^{(j)})/(\theta^{(k)} - \theta^{(i)})$ is given by

$$11X^6 + 33X^5 + 64X^4 + 73X^3 + 64X^2 + 33X + 11$$

from which we calculate the height to be 0.48536.

We have in the previous notation

$$\beta = \pm\theta^{a_1}\pi_1^{s_1}$$

with $H = \max\{|a_1|, s_1\}$. The aim is to find an upper bound on $H$. The first task is to bound $s_1$ in terms of $H$. The polynomial $X^3 - X^2 + X + 1$ has one root in $\mathbb{Q}_{11}$ which we denote by

$$\phi^{(1)} = 6 + 3 \cdot 11 + 8 \cdot 11^2 + O(11^3).$$

The other two roots, which we denote by $\phi^{(2)}$ and $\phi^{(3)}$, are the roots of the following 11-adic quadratic polynomial:

$$X^2 + (5 + 3 \cdot 11 + 8 \cdot 11^2 + O(11^3))X + (9 + 2 \cdot 11 + 11^2 + O(11^3)).$$

The complex roots of $X^3 - X^2 + X + 1$ we denote by $\theta^{(1)}, \theta^{(2)}, \theta^{(3)}$ as usual. As our choice of indices in Lemma VIII.5 we choose $(i_1, j_1, k_1) = (1, 2, 3)$. With this choice we have, to recap,

$$\alpha_1 = \left(\frac{\phi^{(2)} - \phi^{(3)}}{\phi^{(3)} - \phi^{(1)}}\right), \quad \alpha_2 = \left(\frac{\phi^{(1)} - \phi^{(2)}}{\phi^{(3)} - \phi^{(1)}}\right)$$

and

$$\tau_1 = \left(\frac{\phi^{(1)}}{\phi^{(2)}}\right)^{a_1} \left(\frac{\pi_1^{(1)}}{\pi_1^{(2)}}\right)^{s_1}, \quad \tau_2 = \left(\frac{\phi^{(3)}}{\phi^{(2)}}\right)^{a_1} \left(\frac{\pi_1^{(3)}}{\pi_1^{(2)}}\right)^{s_1}$$

We wish to solve the equation

$$\alpha_1\tau_1 + \alpha_2\tau_2 + 1 = 0$$

where the only unknowns are $a_1$ and $s_1$. It is now an easy matter to check the following valuations:

$$\mathrm{ord}_{11}(\alpha_1) = 1/2, \quad \mathrm{ord}_{11}(\alpha_2) = \mathrm{ord}_{11}(\tau_2) = 0, \quad \mathrm{ord}_{11}(\tau_1) = s_1,$$

from which we deduce that

$$s_1 + 1/2 = \mathrm{ord}_{11}(-\alpha_2\tau_2 - 1).$$

But we are now in a position to apply Yu's theorem, see Appendix A.2, to the right hand side of the above equality.

We first need to compute the various modified heights

$$\begin{aligned} h_1 = h_m(\alpha_2) &= \max\left\{0.48536, 0.0383, \frac{\log 11}{6}\right\} = 0.48536, \\ h_2 = h_m\left(\frac{\phi^{(3)}}{\phi^{(2)}}\right) &= \max\left\{0.304688, 0.0313, \frac{\log 11}{6}\right\} = 0.39965, \\ h_3 = h_m\left(\frac{\pi_1^{(3)}}{\pi_1^{(2)}}\right) &= \max\left\{0.830993, 0.03533, \frac{\log 11}{6}\right\} = 0.830993. \end{aligned}$$

We set $H = \max\{h_1, h_2, h_3\} = 0.830993$. If we let $\mathfrak{p}_{11}$ denote a prime ideal of $K^{Gal}$ lying above $\mathfrak{p}_1$, then Yu's theorem tells us that

$$\mathrm{ord}_{\mathfrak{p}_{11}}(-\alpha_2\tau_2 - 1) < c_{18}(\log H + c_{19}),$$

where

$$c_{18} = 30760 \cdot 25^3 \cdot 4^{10} \cdot 11^2 \cdot 12^5\, h_1 h_2 h_3 (\log 11)^{-4} \log(2^{11} \cdot 4^2 \cdot 12^2 H) = 1.124 \cdot 10^{21}$$

and

$$c_{19} = 2\log 12 = 4.9698.$$

Then

$$\begin{aligned} s_1 &= \mathrm{ord}_{11}(-\alpha_2\tau_2 - 1) - 1/2, \\ &= \frac{1}{2}\left(\mathrm{ord}_{\mathfrak{p}_{11}}(-\alpha_2\tau_2 - 1) - 1\right), \\ &< \frac{c_{18}}{2}\left(\log H + c_{19} - 1/c_{18}\right), \\ &= c_6(\log H + c_7). \end{aligned}$$

So we have $c_6 = 5.62 \cdot 10^{20}$ and $c_7 = 4.9698$. If $H = S$ then this gives us that $H \leq 10^{24}$; we therefore need to bound $H$ when $H = A$.

We compute the various other constants needed. In this example we find $c_3 = 1.499769$ and $c_4$ is given by

$$c_4 = \max(\log|\theta^{(1)}|, \log|\theta^{(2)}|) = 0.30468.$$

We then choose $c_5 = 0.07617$. The various constants in Cases A and B we find to be

$$c_{14} = 1.66199\ ,\ c_{15} = c_{16} = 0\ ,\ c_{17} = 0.83010.$$

This leads us to deduce, using the above values for $c_6$ and $c_7$ and the lemma of Pethő and de Weger, that:

**Case A:**

$$H \leq \frac{c_{15} + c_{14}S}{c_4 - (n-1)c_5} = 10.9098S \leq 0.6519 \cdot 10^{24}.$$

**Case B:**

$$H \leq \frac{c_{16} + c_{17}S}{c_4 - c_5} = 3.6327S \leq 0.2251 \cdot 10^{24}.$$

We need then to deduce an upper bound for $H$ in Case C. This is accomplished by using the result of Appendix A.1 to find

$$-0.453746 \cdot 10^{20} \log(3A/2) \leq \log|\Lambda| \leq 1.0985 - c_5 A$$

which leads us to deduce that in Case C

$$A \leq 0.5748 \cdot 10^{23}.$$

**The first 11-adic reduction step.** We now need to apply our methods to reduce the upper bounds on $S$ and $A$. We have

$$\max(S, A) \leq 10^{24}.$$

Using the equality

$$S + 1/2 = s_1 + 1/2 = \operatorname{ord}_{11}(-\alpha_2\tau_2 - 1)$$

we can reduce the bound on $S$ using the $p$-adic approximation lattices considered in Chapter VI. We find

$$\begin{aligned}
\chi_0 &= \log_{11}(\alpha_2) = 2\phi + 10 + 9\phi 11 + (5\phi + 4)11^2 + O(11^3) \\
&= \chi_{0,1} + \chi_{0,2}\phi, \\
\chi_1 &= \log_{11}\left(\frac{\phi^{(3)}}{\phi^{(2)}}\right) = 10\phi + 2 + (3\phi + 1)11 + (6\phi + 2)11^2 + O(11^3) \\
&= \chi_{1,1} + \chi_{1,2}\phi, \\
\chi_2 &= \log_{11}\left(\frac{\pi_1^{(3)}}{\pi_1^{(2)}}\right) = \phi + 9 + 11(4\phi + 4) + 11^2(\phi + 9) + O(11^3) \\
&= \chi_{2,1} + \chi_{2,2}\phi,
\end{aligned}$$

where $\phi = \phi^{(2)}$. Then assuming that $S \geq 1$ we obtain

$$S + 1/2 = \operatorname{ord}_{11}\left(\chi_0 + a_1\chi_1 + s_1\chi_2\right).$$

In the notation of Chapter VI we set $u = 48$ and compute the above 11-adic logarithms to around 55 digits of precision. We then apply the $p$-adic reduction methods discussed earlier. We consider the lattice, $\mathcal{L}$, generated by the columns of the matrix

$$A = \begin{pmatrix} 1 & 0 & 0 & 0 \\ 0 & 1 & 0 & 0 \\ \chi_{1,1}^{\{\mu\}} & \chi_{2,1}^{\{\mu\}} & 11^u & 0 \\ \chi_{1,2}^{\{\mu\}} & \chi_{2,2}^{\{\mu\}} & 0 & 11^u \end{pmatrix}.$$

In this case we compute $A$ to be $(\vec{a_1}, \vec{a_2}, \vec{a_3}, \vec{a_4})$ where

$$\vec{a_1} = \begin{pmatrix} 1 \\ 0 \\ 4916115410847193742284820787638843172935127657327 1 \\ 6222418869760955989569688400165245651683114888794 6 \end{pmatrix},$$

$$\vec{a_2} = \begin{pmatrix} 0 \\ 0 \\ 4424662732774721936373254116291811841466803907687 7 \\ 8163520410563902932837457119336611977192135677361 5 \end{pmatrix},$$

and $\vec{a_3} = (0, 0, 11^{48}, 0)^t$, $\vec{a_4} = (0, 0, 0, 11^{48})^t$. We then apply the LLL–algorithm to find that a reduced basis is given by $B = (\vec{b_1}, \vec{b_2}, \vec{b_3}, \vec{b_4})$ where

$$\vec{b_1} = \begin{pmatrix} -3206195076262832992096 74 \\ 6349275212839218469100069 \\ -6793882650941251298438996 \\ 3866306976174322845246 46 \end{pmatrix},$$

$$\vec{b_2} = \begin{pmatrix} 3772106613893795877168360 \\ 2046834026897093540976666 \\ 6672278208843796053001676 \\ 5726157940472902813514913 \end{pmatrix},$$

$$\vec{b_3} = \begin{pmatrix} -5031195905649157044756 63 \\ 2401239200282808790628 28 \\ -4783834428241165932396938 \\ 9135891617987144859763079 \end{pmatrix},$$

$$\vec{b_4} = \begin{pmatrix} 1303042897945565360430139 1 \\ -1355563419635373431473 5 \\ -2120949474251310659066433 \\ -1318621050704023258089929 \end{pmatrix}.$$

If we then set

$$\vec{y} = \begin{pmatrix} 0 \\ 0 \\ 1038335831737266425232335084935257655953428545176 5 \\ 7298457587533909940784691569974322561561720058793 5 \end{pmatrix}$$

we deduce, using Lemma V.10, that

$$\ell(\mathcal{L}, \vec{y}) \geq 0.498 \cdot 10^{25}.$$

Hence, by Lemma VI.6 we find that $S \leq 48$. Then we need to look at how this effects the bounds for $H$ in all our cases.

**The first real reduction step.** If $H = S$ then clearly $A \leq H \leq 48$, so we need to look at when $H = A$. This case we divided into three subcases which become:

**Case A:** $H \leq 10.9098S < 524$.
**Case B:** $H \leq 3.6327S < 174$.
**Case C:** In this case we deduce no information about a new bound on $H$ from the new bound on $S$. However, we can obtain a new bound on $H$ by using the linear form in logarithms $\Lambda$ which satisfies the inequality

$$|\Lambda| \leq 3e^{-c_5 A}.$$

So we proceed much as when we dealt with a Thue equation in the last chapter. We need to consider two possible subcases, according to which choice of $i$ satisfies

$$|\beta^{(i)}| = \min_{1 \leq l \leq 2} |\beta^{(l)}|.$$

We first deal with the case $i = 1$. In this case we obtain the linear form

$$\Lambda = \log\left(\frac{\theta^{(1)} - \theta^{(2)}}{\theta^{(3)} - \theta^{(1)}}\right) + a_1 \log\left(\frac{\theta^{(3)}}{\theta^{(2)}}\right) + s_1 \log\left(\frac{\pi^{(3)}}{\pi^{(2)}}\right) + a_0 2\pi\sqrt{-1},$$

with $a_0 \in \mathbb{Z}$ such that $|a_0| \leq 3H$. We find that all three logarithms above have zero real part so we approximate this linear form using the lattice, $\mathcal{L}$, generated by the columns of the matrix

$$A = \begin{pmatrix} 1 & 0 & 0 \\ 0 & 1 & 0 \\ v_1 & v_2 & v_3 \end{pmatrix}$$

where, with $C = 10^{77}$,

$$\begin{aligned} v_1 &= \left[\Im\left(C \log\left(\frac{\theta^{(3)}}{\theta^{(2)}}\right)\right)\right], \\ v_2 &= \left[\Im\left(C \log\left(\frac{\pi^{(3)}}{\pi^{(2)}}\right)\right)\right], \\ v_3 &= [C2\pi]. \end{aligned}$$

We find that an LLL–reduced basis of the lattice $\mathcal{L}$ is given by

$$B = \begin{pmatrix} 45813986072140905459481417 & 16561089972607042058472889 & -47950255292186937749047298 \\ 58192974024359721053091467 & -89311838016919297734723468 & 20847718212569027398167960 \\ -49881490689096800691710630 & -12757978463024807747173142 & -75984441016165974379745690 \end{pmatrix}$$

Then, by Lemma V.10, we deduce that

$$\ell(\mathcal{L}, \vec{y}) \geq 0.331 \cdot 10^{26}$$

where $\vec{y} = (0, 0, v_0)^t$, with

$$v_0 = \left[\Im\left(\log\left(\frac{\theta^{(1)} - \theta^{(2)}}{\theta^{(3)} - \theta^{(1)}}\right)\right)\right].$$

Using Lemma VI.1 we can then deduce that in Case C with $i = 1$ we have $H \leq 1571$.

We now have to consider the case $i = 2$. Now we obtain the linear form

$$\Lambda = \log\left(\frac{\theta^{(2)} - \theta^{(1)}}{\theta^{(3)} - \theta^{(2)}}\right) + a_1 \log\left(\frac{\theta^{(3)}}{\theta^{(1)}}\right) + s_1 \log\left(\frac{\pi^{(3)}}{\pi^{(1)}}\right) + a_0 2\pi\sqrt{-1},$$

again with $a_0 \in \mathbb{Z}$ such that $|a_0| \leq 3H$. We find that all three logarithms above all have non-zero real and imaginary parts so we approximate this linear form using the lattice, $\mathcal{L}$, generated by the columns of the matrix

$$A = \begin{pmatrix} 1 & 0 & 0 \\ u_1 & u_2 & 0 \\ v_1 & v_2 & v_3 \end{pmatrix}$$

where, with $C = 10^{38}$,

$$u_1 = \left[\Re\left(C \log\left(\tfrac{\theta^{(3)}}{\theta^{(1)}}\right)\right)\right], \quad u_2 = \left[\Re\left(C \log\left(\tfrac{\pi^{(3)}}{\pi^{(1)}}\right)\right)\right],$$

$$v_1 = \left[\Im\left(C \log\left(\tfrac{\theta^{(3)}}{\theta^{(1)}}\right)\right)\right],$$

$$v_2 = \left[\Im\left(C \log\left(\tfrac{\pi^{(3)}}{\pi^{(1)}}\right)\right)\right], \quad v_3 = [C2\pi]$$

We find that an LLL–reduced basis of the lattice $\mathcal{L}$ is given by

$$B = \begin{pmatrix} 861978669403938973839563 1 & 29108622913049408578940 20 & 201890912292581046100733 72 \\ -844088055425244152738943 0 & 34252610568227263545853 52 & 133932010945965461283880 20 \\ 324698280699519288846266 4 & 21385013973943592893471555 & 437637230535181865708807 9 \end{pmatrix}.$$

Then, by Lemma V.10, we deduce that

$$\ell(\mathcal{L}, \vec{y}) \geq 0.51 \cdot 10^{25}$$

where $\vec{y} = (0, u_0, v_0)^t$, with

$$\begin{aligned} u_0 &= \left[\Re\left(\log\left(\frac{\theta^{(2)} - \theta^{(1)}}{\theta^{(3)} - \theta^{(2)}}\right)\right)\right], \\ v_0 &= \left[\Im\left(\log\left(\frac{\theta^{(2)} - \theta^{(1)}}{\theta^{(3)} - \theta^{(2)}}\right)\right)\right]. \end{aligned}$$

Using Lemma VI.2 we can then deduce that in Case C with $i = 1$ we have $H \leq 430$.

So in all cases after one 11-adic reduction step and one real reduction step we can conclude $S \leq 48$ and $H \leq 1571$. These are still rather large, although now in the range of explicit enumeration. We can make them even smaller by performing further reductions which we shall now do.

**The second 11-adic reduction step.** Using our new bound on $H$ of 1571 we can further reduce the upper bound on $S$ by this time setting $u = 8$ and applying the 11-adic reduction step again. This time we are led to consider the lattice, $\mathcal{L}$, generated by the columns of the matrix

$$A = \begin{pmatrix} 1 & 0 & 0 & 0 \\ 0 & 1 & 0 & 0 \\ 84075532 & 181482992 & 214358881 & 0 \\ 146589123 & 18592905 & 0 & 21435888 \end{pmatrix}.$$

We then apply the LLL–algorithm to find that a reduced basis is given by

$$B = \begin{pmatrix} -3396 & 5817 & -8911 & 4412 \\ -215 & -2409 & -4545 & -49127 \\ -4118 & -7286 & -1109 & 13622 \\ 4138 & -2143 & -8350 & 14705 \end{pmatrix}.$$

Then from Lemma V.10 we find that

$$\ell(\mathcal{L}, \vec{y}) \geq 2881$$

where $y = (0, 0, 99676773, 88284516)^t$. And so, by Lemma VI.6 we find that $S \leq 8$.

**The second real reduction step.** We then have to determine how this new small upper bound on $S$ affects the upper bounds on $H$. We may assume that $H = A$ and divide the discussion again into the three subcases

**Case A:** $H \leq 10.9098S < 87$.
**Case B:** $H \leq 3.6327S < 29$.
**Case C:** Again Case C becomes the difficult case as we need to reduce a linear form in logarithms. There are two cases according to whether $i = 1$ or $i = 2$. In the first we deduce that $H \leq 277$, using $C = 10^{13}$, whilst in the second we deduce that $H \leq 102$, using $C = 10^7$.

We find that applying the real reduction step again in the case $i = 1$ with the new bound of 277 leads us to deduce in this case that $H \leq 221$, using $C = 10^{10}$. We were unable to reduce the bounds in either the 11-adic or the real cases any further and so we have to settle with the final bounds of

$$S \leq 8\ ,\ A \leq 221.$$

So all that remains is to go through all the 9 possible values for $s_1$ and the 443 values for $a_1$.

In the case under consideration with, $s_2 = 0$, we find that there are the following solutions to

$$X^3 - X^2Y + XY^2 + Y^3 = \pm 11^s$$

with $Y \geq 0$.

| $s_1$ | $a_1$ | X | Y |
|---|---|---|---|
| 0 | 0 | 1 | 0 |
| 0 | 1 | 0 | 1 |
| 0 | 4 | −1 | 2 |
| 0 | 17 | −56 | 103 |
| 1 | 0 | 1 | 2 |
| 2 | 1 | −4 | 3 |

## VIII.5. Exercises

1). In Lemma VIII.5 show that there is always a choice of indices such that

$$\mathrm{ord}_{p_l}(\alpha_2) \geq 0\ ,\ \mathrm{ord}_{p_l}\left(\frac{\pi_l^{(k)}}{\pi_l^{(j)}}\right) = 0 \text{ and } \mathrm{ord}_{p_l}\left(\frac{\pi_l^{(i)}}{\pi_l^{(j)}}\right) = h_l.$$

2). Prove that there is at most one prime ideal $\mathfrak{p}_i$ dividing $p$ with

$$\mathrm{ord}_{\mathfrak{p}_i}(X - \theta Y) > \frac{\max\{e_1, \ldots, e_m\}}{2}\mathrm{ord}_p D(\theta),$$

and that such a $\mathfrak{p}_i$ must satisfy $e_i = f_i = 1$.

3). Fill in all the gaps and the computations in the example of a Thue–Mahler equation above. Do not forget to do the case $s_2 = 1$.

CHAPTER IX

# $S$-unit equations

In the preceding chapters we have seen how Thue and Thue–Mahler equations can be reduced to the study of equations of the form

$$\alpha_1\tau_1 + \alpha_2\tau_2 + 1 = 0$$

where the $\tau_i$ are allowed to range over two finitely generated subgroups of the algebraic numbers. These subgroups were determined by the equation we wished to solve. Such two term equations can be studied in their own right, a topic to which we shall devote this chapter. We shall see, in this chapter and the next, how we can reduce the study of other types of diophantine equations to the study of such two term equations. We first derive upper bounds on the solutions of such equations using the theory of linear forms in logarithms. We shall then reduce these bounds using approximation lattices, just as we did for Thue and Thue–Mahler equations. The main problem then is to locate all the 'small' solutions. To solve this remaining problem we shall introduce a sieving technique which is very efficient in practice and lends itself to implementation either on a parallel computer or in a distributed environment. We shall end the chapter with a discussion of the possible applications of such equations.

It is perhaps worth mentioning here that recently Wildanger [**216**] has given a very fast method to find all the small solutions to the equation above when $\tau_1, \tau_2 \in \mathcal{O}_K^*$. We shall not cover Wildanger's method except to note that it works substantially faster than the sieving strategy mentioned below. The method of Wildanger can also be generalized to the equations considered in this chapter [**184**].

## IX.1. $S$-unit equations

Let $H_1$ and $H_2$ be two finitely generated multiplicative subgroups of the algebraic numbers and assume that the generators of these two groups are explicitly given to us. Also let $\alpha_1$ and $\alpha_2$ denote two fixed non-zero algebraic numbers. In this chapter we wish to give a practical method to locate all the solutions to the equation

$$\alpha_1\tau_1 + \alpha_2\tau_2 + 1 = 0 \text{ with } (\tau_1, \tau_2) \in H_1 \times H_2. \tag{IX.1}$$

In what follows we shall assume that $H_1$ and $H_2$ are both torsion free. This is no problem in practice as if they do contain torsion elements we can just increase the number of cases for $(\alpha_1, \alpha_2)$ and remove the torsion from $H_1$ and $H_2$. We shall let $K$ denote the field of definition of $H_1, H_2, \alpha_1$ and $\alpha_2$, i.e. $H_1$

and $H_2$ are subgroups of $K^*$ and $\alpha_1, \alpha_2$ are elements of $K^*$. We shall assume (although it is not necessary) that $K$ is the smallest such field. Clearly $K$ is a field of finite degree over $\mathbb{Q}$.

We shall let $S_1$ and $S_2$ denote the set of primes (places), both finite and infinite, in the support of the groups $H_1$ and $H_2$ respectively. In other words

$$S_i = \{\mathfrak{p} \in M_K : |\alpha|_{\mathfrak{p}} \neq 1 \text{ for some } \alpha \in H_i\}.$$

Hence equation (IX.1) is a special case of the equation where $\tau_1$ and $\tau_2$ range over the group of $S$-units of $K$ (where $S = S_1 \cup S_2$). Because of this we shall often refer to equation (IX.1) as a two term $S$-unit equation.

That an $S$-unit equation has only finitely many solutions was proved by Siegel [**168**]. Coates gave an effective proof in the special case where the $S$-unit equation arises from a Thue–Mahler equation. However, Győry [**90**] was the first to give explicit effective upper bounds on the heights of the solutions of such an equation. Evertse [**53**] has shown that the equation

$$\alpha_1\tau_1 + \alpha_2\tau_2 = 1$$

where $\tau_i \in \mathcal{O}_S^* \subset K^*$ has at most

$$3 \cdot 7^{[K:\mathbb{Q}]+2\#S}$$

solutions.

We shall let $r_i$ denote the rank of the group $H_i$ and then define $t_i$ by $t_i = |S_i| - 1$. Then one can define a lattice in $\mathbb{R}^{t_i}$, associated to the group $H_i$, via the map

$$Log_i : \begin{cases} H_i & \longrightarrow & \mathbb{R}^{t_i} \\ \alpha & \longmapsto & (\log|\alpha|_{\mathfrak{p}_1}, \dots, \log|\alpha|_{\mathfrak{p}_{t_i}})^t \end{cases}$$

The components in the image vector come from the elements of $S_i$, the missing one corresponding by convention to the $(r+1)$th-embedding of $K$, where $r$ is the unit rank of $K$. The image of $Log_i$ is a lattice of rank $r_i$. Before we continue we note that what we really want to do is express each solution, $(\tau_1, \tau_2)$, in terms of the generators of $H_1 \times H_2$ and then solve for the (exponential) indeterminates. For example suppose that

$$H_1 = \langle\beta_{1,1}\rangle \times \cdots \times \langle\beta_{r_1,1}\rangle \text{ and } H_2 = \langle\beta_{1,2}\rangle \times \cdots \times \langle\beta_{r_1,2}\rangle,$$

then we wish to find the variables $a_{i,j}$ when we express $\tau_1$ and $\tau_2$ as

$$\tau_1 = \prod_{i=1}^{r_1} \beta_{i,1}^{a_{i,1}} \text{ and } \tau_2 = \prod_{i=1}^{r_2} \beta_{i,2}^{a_{i,2}}.$$

Clearly different choices of generators may lead to larger/smaller values of the $a_{i,j}$. We would like the $a_{i,j}$ to be as small as possible. So we need to decide what is a good choice of generators for the groups $H_i$. To aid us we must first give a general piece of folklore, which even if not true at least acts as a very good rule of thumb in most cases.

**Folklore.** If a diophantine equation has only finitely many solutions then those solutions are small in 'height' when compared to the parameters of the equation.

This folklore is, however, only widely believed because of the large amount of experimental evidence which now exists to support it. This is an important argument to support work on practical solution methods for diophantine equations. Apparently before such work was carried out many specialists believed that diophantine equations usually have some small and some large solutions, these phantom large solutions being the reason why the transcendence results produced such large upper bounds.

Hence in the above we expect our $\tau_i$ to have small height. Now if the $a_{i,j}$ are also to be small then the generators must also have small height and be 'orthogonal' is some vague sense. But we have some freedom in choosing the generators; in fact the generators of $H_i$ are only unique up to a $GL_{r_i}(\mathbb{Z})$-transformation. This is equivalent to choosing a new basis for the lattice which is the image of the map $Log_i$. It turns out that a sensible thing to do is to choose the generators $\beta_{i,j}$ such that the image of these generators under $Log_i$ is an LLL–reduced basis. It is clear that given a set of generators we can always transform it to a new set of generators so that the image of the new generators is LLL–reduced. Hence this is not a problem in practice but should be born in mind.

As in the case of Thue and Thue–Mahler equations we wish to bound

$$A = \max_{i,j} |a_{i,j}|.$$

An upper bound on $A$ will certainly give us only a finite number of cases to check. Say we have obtained an upper bound, $B$. Then this 'only' leaves

$$(2B)^{r_1+r_2}$$

cases to check, from which it is clear that even if $B$ is very small we could have a large number of cases to check. For the rest of this section we shall consider the problem of deducing a 'small' upper bound, $B$, on $A$. Then in the next section we shall show how one can use a sieving technique to find the solutions below this small upper bound.

We choose $b$ and $\mathfrak{p}_h \in S_b$ such that $A = |a_{k,b}|$ for some $k$ and

$$|\tau_b|_{\mathfrak{p}_h} = \min_{\mathfrak{p}_i \in S_b} |\tau_b|_{\mathfrak{p}_i}.$$

At this stage we do not know the exact values of $b$ and $\mathfrak{p}_h$ and hence we shall need to perform the following for all possible values, much as we looped through all values of the index $i$ in the earlier discussion of Thue–Mahler equations.

Let $C \in \mathbb{R}^{t_b \times r_b}$ denote the matrix whose columns are the image of the generators of $H_b$ under $Log_b$. Let $C'$ denote any choice of $r_b$ rows from $C$

which gives a non-zero determinant. Such a choice exists as $H_b$ is a group of rank $r_b$. We set

$$c_1 = \|C'^{-1}\|_\infty,$$

where $\|.\|_\infty$ denotes the infinity norm of a matrix, i.e. the row sum norm. We set $c_2 = 1/c_1$ and choose $c_3$ to be any constant such that

$$0 < c_3 < c_2/t_b.$$

From the analysis we carry out below the best choice for $c_3$ is one which is very near the top end of this possible range, i.e. around $0.99c_2/t_b$. We shall now show that the value of $|\tau_b|_{\mathfrak{p}_h}$, which we have already chosen to be small, will in fact be exponentially small in terms of $A$.

LEMMA IX.1. *With the conventions denoted above we have*

$$|\tau_b|_{\mathfrak{p}_h} \le e^{-c_3 A}.$$

PROOF. Let $U = \{u_1, \dots, u_{r_b}\}$ denote the choice of rows of $C$ which make up the matrix $C'$ above. We then have the matrix equation

$$\begin{pmatrix} a_{1,b} \\ \dots \\ a_{r_b,b} \end{pmatrix} = C'^{-1} \begin{pmatrix} \log|\tau_b|_{\mathfrak{p}_{u_1}} \\ \dots \\ \log|\tau_b|_{\mathfrak{p}_{u_b}} \end{pmatrix}.$$

Then taking the infinity matrix norm of both sides we obtain

$$A \le \|C'^{-1}\|_\infty \max_{u\in U}(\log|\tau_b|_{\mathfrak{p}_u}) \le c_1 \max_{\mathfrak{p}\in S_b}(\log|\tau_b|_{\mathfrak{p}}).$$

We let $\mathfrak{p}_g$ denote the element of $S_b$ such that

$$\log|\tau_b|_{\mathfrak{p}_g} = \max_{\mathfrak{p}\in S_b}(\log|\tau_b|_{\mathfrak{p}}).$$

Then there are two cases to consider, either

$$|\tau_b|_{\mathfrak{p}_g} \ge e^{c_2 A} \quad \text{or} \quad |\tau_b|_{\mathfrak{p}_g} \le e^{-c_2 A}.$$

We shall assume that $|\tau_b|_{\mathfrak{p}_h} > e^{-c_3 A}$ and in both of these cases try to deduce a contradiction:

**Case 1:** $|\tau_b|_{\mathfrak{p}_g} \ge e^{c_2 A}$. In this case we have

$$\prod_{\mathfrak{p}\in S_b} |\tau_b|_{\mathfrak{p}} = 1$$

and so

$$e^{c_2 A} \le |\tau_b|_{\mathfrak{p}_g} = \prod_{\mathfrak{p}\in S_b, \mathfrak{p}\ne\mathfrak{p}_g} |\tau_b|_{\mathfrak{p}}^{-1} < e^{t_b c_3 A}.$$

Hence as $(c_2 - t_b c_3)A < 0$ we must have $A < 0$ which is a contradiction.

**Case 2:** $|\tau_b|_{\mathfrak{p}_g} \le e^{-c_2 A}$. This case is even easier as we have

$$e^{-c_3 A} < |\tau_b|_{\mathfrak{p}_h} \le |\tau_b|_{\mathfrak{p}_g} \le e^{-c_2 A}.$$

Hence $(c_2 - c_3)A < 0$ and we again conclude that $A < 0$. □

Note you should compare the above lemma with Lemma VII.2 and what we called cases $A$ and $B$ when looking at Thue and Thue–Mahler equations. The crucial point is that in the above lemma we are treating both the finite and infinite places on the same footing. This leads to a much more efficient algorithm when prime ideals of degree greater than one lie in the support of our two groups $H_1$ and $H_2$.

We now have to turn the fact that we know that one valuation of $\tau_b$ is exponentially small into deducing that a linear form in logarithms is also small. Then combining estimates from the theory of linear forms in logarithms with the above inequality will give us an upper bound on $A$ which we can hope to reduce using approximation lattices.

We set $c_4 = |\alpha_b|_{\mathfrak{p}_h}$, $d = 3-b$, $\Lambda_b = \alpha_b\tau_b = -\alpha_d\tau_d - 1 = -\Lambda_d - 1$ and divide the discussion into two cases: either $\mathfrak{p}_h$ corresponds to a finite prime ideal or it corresponds to an infinite place. We shall treat this latter case first.

**IX.1.1. $\mathfrak{p}_h$ is an infinite real place.** By abuse of notation we let $h$ also denote the conjugate of $K$ which corresponds to the place $\mathfrak{p}_h$, i.e. $|\alpha|_{\mathfrak{p}_h} = |\alpha^{(h)}|$. We set $c_5 = \log(2c_4)/c_3$, then if $A \geq c_5$ we have

$$\begin{aligned} |-\Lambda_d^{(h)} - 1| &= |\Lambda_b^{(h)}| = |\Lambda_b|_{\mathfrak{p}_h}, \\ &\leq c_4 e^{-c_3 A} \leq c_4 e^{-c_3 c_5}, \\ &= c_4 e^{-\log(2c_4)} = \frac{1}{2}. \end{aligned}$$

Hence by Lemma B.2 we have

$$|\log|\Lambda_d^{(h)}|| \leq |\log(-\Lambda_d^{(h)})| \leq c_6 e^{-c_7 A}$$

where $c_6 = 2c_4$ and $c_7 = c_3$. But the left hand side of this inequality is equal to

$$\left|\log|\alpha_d^{(h)}| + \sum_{i=1}^{r_d} a_{i,d}\log|\beta_{i,d}^{(h)}|\right|.$$

Now we can find a lower bound on this linear form in logarithms using the result of Baker and Wüstholz, Appendix A.1, i.e. we can compute an explicit constant, $c_8$, such that

$$-c_8\log(A) < \log|\log|\Lambda_d^{(h)}|| \leq \log(c_6) - c_7 A.$$

Hence

$$A < \frac{1}{c_7}\left(\log(c_6) + c_8\log A\right).$$

The right hand side of this last inequality grows asymptotically slower than the left hand side so this must lead to an upper bound on $A$. An explicit upper bound can be found using the lemma of Pethő and de Weger, Appendix B, Lemma B.1 We find

$$A \leq \frac{2}{c_7}\left(\log(c_6) + c_8\log(c_8/c_7)\right) = A_1.$$

The constant $A_1$ will usually turn out to be rather large. However, we can use the inequality

$$\left|\log|\alpha_d^{(h)}| + \sum_{i=1}^{r_d} a_{i,d} \log|\beta_{i,d}^{(h)}|\right| \le c_6 e^{-c_7 A}$$

and the reduction techniques of Section VI.3 to find a 'small' upper bound on $A$ (provided $A \ge c_5$).

**IX.1.2. $\mathfrak{p}_h$ is an infinite complex place.** Again by abuse of notation we let $h$ also denote the conjugate of algebraic numbers which corresponds to the place $\mathfrak{p}_h$, i.e. $|\alpha|_{\mathfrak{p}_h} = |\alpha^{(h)}|^2$. To show the analogy with the real case we shall keep the same numbering for analogous constants (we hope this does not confuse you). This time we set

$$c_5 = 2\log(2\sqrt{c_4})/c_3,$$

then $A \ge c_5$ implies

$$\begin{aligned} |-\Lambda_d^{(h)} - 1| &= |\Lambda_b^{(h)}| = |\Lambda_b|_{\mathfrak{p}_h}^{1/2}, \\ &\le \sqrt{c_4}e^{-c_3 A/2} \le \sqrt{c_4}e^{-c_3 c_5/2}, \\ &= \sqrt{c_4}e^{-\log(2\sqrt{c_4})} = \frac{1}{2}. \end{aligned}$$

Again using Lemma B.2 we have

$$|\log(-\Lambda_d^{(h)})| \le c_6 e^{-c_7 A}$$

where this time $c_6 = 2\sqrt{c_4}$ and $c_7 = c_3/2$. But the left hand side of this inequality is equal to

$$\left|\log(-\alpha_d^{(h)}) + \sum_{i=1}^{r_d} a_{i,d} \log \beta_{i,d}^{(h)} + a_{0,d} 2\pi\sqrt{-1}\right|$$

with $|a_{0,d}| \le (r_d + 1)A$. We can find a lower bound on this linear form in logarithms using the result of Baker and Wüstholz again,

$$-c_8 \log((r_d + 1)A) < \log|\log(-\Lambda_d^{(h)})| \le \log(c_6) - c_7 A,$$

for some explicit constant $c_8$. Hence

$$A < \frac{1}{c_7}\left(\log(c_6) + c_8 \log((r_d + 1)A)\right).$$

The lemma of Pethő and de Weger then gives us

$$A < \frac{2}{c_7}\left(\log(c_6) + c_8 \log(r_d + 1) + c_8 \log(c_8/c_7)\right) = A_2.$$

However, we can now use the inequality

$$\left|\log(-\alpha_d^{(h)}) + \sum_{i=1}^{r_d} a_{i,d} \log \beta_{i,d}^{(h)} + a_{0,d} 2\pi\sqrt{-1}\right| \le c_6 e^{-c_7 A}$$

and the reduction techniques of Section VI.3 to find a 'small' upper bound on $A$, (provided again $A \geq c_5$).

**IX.1.3. $\mathfrak{p}_h$ is a finite place.** We write $f_h$ for the residue degree of $\mathfrak{p}_h$, $e_h$ for the ramification index and $p_h$ for the rational prime lying below $\mathfrak{p}_h$. Our inequality

$$|\Lambda_b|_{\mathfrak{p}_h} \leq c_4 e^{-c_3 A}$$

can then be written as

$$-f_h \text{ord}_{\mathfrak{p}_h}(\Lambda_b) \log(p_h) \leq -c_3 A + \log c_4.$$

Hence

$$\text{ord}_{\mathfrak{p}_h}(\Lambda_b) \geq (c_3 A - \log c_4)/(f_h \log p_h) = e_h(c_9 A - c_{10})$$

where $c_9$ and $c_{10}$ have obvious definitions. Then if $A > c_{10}/c_9$ then $\text{ord}_{\mathfrak{p}_h}(\Lambda_b) > 0$ and so $\text{ord}_{\mathfrak{p}_h}(\Lambda_d) = 0$, which leads us to deduce:

LEMMA IX.2. *If by some miracle we know a priori that*

$$\text{ord}_{\mathfrak{p}_h}(\alpha_1 \tau_1) = \text{ord}_{\mathfrak{p}_h}(\alpha_2 \tau_2)$$

*then* $A \leq c_{10}/c_9$.

In many examples that one encounters the above miracle often occurs, owing to some Galois action that we can exploit, hence it is not as special as might appear at first sight. In addition the upper bound we obtain is very, very small and hence no further reduction needs to be performed for this place. We now proceed to consider the cases when we are not so lucky.

LEMMA IX.3. *If $A > c_{10}/c_9$ then there exists a finite set of possible elements $\mu_i \in K$ for $i = 0, \ldots, s_d$ such that*

1. $\text{ord}_{\mathfrak{p}_h}(\mu_i) = 0$.
2. $s_d = r_d$ *if* $\mathfrak{p}_h \notin S_d$ *otherwise* $s_d = r_d - 1$.
3. *There are integers $b_{i,d}$, with $|b_{i,d}| \leq |a_{i,d}| \leq A$, such that*

$$-\Lambda_d = \mu_0 \prod_{i=1}^{s_d} \mu_i^{b_{i,d}}.$$

PROOF. If $\mathfrak{p}_h \notin S_d$ then we must have $\text{ord}_{\mathfrak{p}_h}(\alpha_d) = 0$ and we can take $\mu_0 = \alpha_0$ and for $i = 1, \ldots, s_d$ we can take $\mu_i = \beta_{i,d}$. Hence we shall assume that $\mathfrak{p}_h \in S_d$. For $i = 1, \ldots, r_d$ set $n_i = \text{ord}_{\mathfrak{p}_h}(\beta_{i,d})$ and set $n_0 = \text{ord}_{\mathfrak{p}_h}(\alpha_d)$. We then define $k \in \{1, \ldots, r_d\}$ to be an index which satisfies

$$n_k = \min_{1 \leq i \leq r_d, n_i \neq 0} |n_i|,$$

such an index exists by the assumption that $\mathfrak{p}_h \in S_d$. By relabelling the $\beta_{i,d}$ we can assume that $k = r_d$. Now as $A > c_{10}/c_9$ we know that $\text{ord}_{\mathfrak{p}_h}(\alpha_d \tau_d) = 0$ and so we have the linear equation

$$n_0 + \sum_{i=1}^{r_d} a_{i,d} n_i = 0.$$

We define $b_{i,d}$ and $r_i$ by the equation $a_{i,d} = n_{r_d} b_{i,d} + r_i$ with $0 \leq r_i < |n_{r_d}|$. Hence $|b_{i,d}| \leq |a_{i,d}| \leq A$. We define

$$\mu_i = \beta_{i,d}^{n_{r_d}} \beta_{r_d,d}^{-n_i}$$

for $i \in \{1, \ldots, r_d - 1\}$, it is clear then that $\mathrm{ord}_{\mathfrak{p}_h}(\mu_i) = 0$ for $i = 1, \ldots, r_d - 1$. We need therefore only find a $\mu_0 \in K$ which satisfies the conditions of the lemma. If we set

$$\sigma = -\left(n_0 + \sum_{i=1}^{r_d-1} n_i r_i\right)$$

then we have

$$\begin{aligned} \mu_0 &= -\alpha_d \prod_{i=1}^{r_d} \beta_{i,d}^{a_{i,d}} \prod_{i=1}^{r_d-1} \mu_i^{-b_{i,d}}, \\ &= -\alpha_d \beta_{r_d,d}^{a_{r_d,d}} \prod_{i=1}^{r_d-1} \left(\beta_{i,d}^{a_{i,d} - n_{r_d} b_{i,d}} \beta_{r_d,d}^{n_i b_{i,d}}\right), \\ &= -\alpha_d \beta_{r_d,d}^{a_{r_d,d} + \sum_{i=1}^{r_d-1} n_i b_{i,d}} \prod_{i=1}^{r_d-1} \beta_{i,d}^{r_i}. \end{aligned}$$

If we set

$$\phi = a_{r_d,d} + \sum_{i=1}^{r_d-1} n_i b_{i,d}$$

then, as we require $\mathrm{ord}_{\mathfrak{p}_h}(\mu_0) = 0$, we must have

$$n_0 + n_{r_d}\phi + \sum_{i=1}^{r_d-1} n_i r_i = 0$$

i.e. $n_{r_d}\phi = \sigma$. Hence we set

$$\mu_0 = -\alpha_d \beta_{r_d,d}^{\sigma/n_{r_d}} \prod_{i=1}^{r_d-1} \beta_{i,d}^{r_i}.$$

Therefore if we want $\mu_0 \in K$ we need only check if $\beta_{r_d,d}^{\sigma/n_{r_d}} \in K$. This will certainly be true if $\sigma \equiv 0 \pmod{n_{r_d}}$ (which will also give us a restriction on the possible values of $r_i$). But we notice that

$$\begin{aligned} \sigma &= -\left(n_0 + \sum_{i=1}^{r_d-1} n_i (a_{i,d} - n_{r_d} b_{i,d})\right), \\ &= \left(n_{r_d} a_{r_d,d} + \sum_{i=1}^{r_d-1} n_{r_d} b_{i,d}\right), \\ &\equiv 0 \pmod{n_{r_d}}. \end{aligned}$$

So we always have $\mu_0 \in K$. □

We can now apply Yu's theorem, Appendix A.2, to find two constants, $c_{11}$ and $c_{12}$, such that

$$\mathrm{ord}_{\mathfrak{p}_h}(\Lambda_b) = \mathrm{ord}_{\mathfrak{p}_h}\left(\mu_0 \prod_{i=1}^{s_d} \mu_i^{b_{i,d}} - 1\right) \leq c_{11} \log A + c_{12}.$$

Combining this inequality with the above lower bound for $\mathrm{ord}_{\mathfrak{p}_h}(\Lambda_b)$ we find

$$e_h c_9 A < c_{11} \log A + c_{12} + e_h c_{10}.$$

Now the left hand side of this inequality grows asymptotically faster than the right hand side, so this means that $A$ is bounded. Again an explicit bound can be found using the lemma of Pethő and de Weger:

$$A \leq \frac{2}{e_h c_9}\left(c_{12} + e_h c_{10} + c_{11} \log(c_1 1/e_h c_9)\right) = A_3.$$

We wish to reduce this upper bound on $A$ to something much smaller. To do this we shall use the $p$-adic reduction methods of VI.4. However, these considered only linear forms and we therefore need to convert the quantity

$$\Lambda_b = \mu_0 \prod_{i=1}^{s_d} \mu_i^{b_{i,d}} - 1$$

into a linear form.

This is done by using $p$-adic logarithms, where $p$ is the rational prime lying below the prime ideal $\mathfrak{p}_h$. If $\mathrm{ord}_p(\Lambda_b) \geq 1/(p-1)$ then by Lemma II.9 we have

$$\mathrm{ord}_p(\Lambda_b) = \mathrm{ord}_p\left(\log_p \mu_0 + \sum_{i=1}^{r_d - 1} b_{i,d} \log_p \mu_i\right) > c_9 A - c_{10},$$

which is exactly the situation considered in Section VI.4. Therefore we need only ensure that $\mathrm{ord}_p(\Lambda_b) \geq 1/(p-1)$ but this will be true when $A > (1 + c_{10})/c_9$ as then

$$\mathrm{ord}_p(\Lambda_b) > c_9 A - c_{10} > 1.$$

## IX.2. Sieving

So using the method of the last section we know that an $S$-unit equation has only finitely many solutions. We also know we can bound the exponential variables in such an equation by a 'small' upper bound. By 'small' what we actually mean is small in comparison to the original bounds we obtained by applying methods from transcendence theory.

Suppose the upper bound is $B$. Then the number of possible solutions left is $O(B^{r_1+r_2})$, which can be rather large. We therefore need a much more efficient way of locating these solutions than a naive brute force search. One method is to sieve using a set of prime ideals.

We shall first explain how this works with prime ideals of fixed norm and then we shall explain how to dovetail a sieve together using a set of prime ideals of different norms.

**Sieving with prime ideals of the same norm.** Let $\mathfrak{p}_i$, for $i = 1, \dots, s$, denote prime ideals of $K$ of the same norm which are not in the support of the groups $H_1$ and $H_2$, i.e. $\mathfrak{p}_i \notin S_1 \cup S_2$. We shall also assume that the $\mathfrak{p}_i$ do not lie in the support of $\alpha_1$ or $\alpha_2$. This precludes a finite number of prime ideals from our method. In addition we shall tacitly assume that the common norm is 'small'.

Suppose the $\mathfrak{p}_i$ lie above the rational prime $p$ and have residue degree $f$. Set $q = p^f$ and let $\phi_i$ denote the 'reduction mod $\mathfrak{p}_i$' map

$$\phi_i : K^* \to K^*_{\mathfrak{p}_i}/(\mathfrak{p}_i) \cong \mathbb{F}_q^*.$$

Here we will need to make an explicit choice of isomorphism, any choice will do. Our eventual aim is to solve the equation

$$\alpha_1 \prod_{i=1}^{r_1} \beta_{i,1}^{a_{i,1}} + \alpha_2 \prod_{i=1}^{r_2} \beta_{i,2}^{a_{i,2}} + 1 = 0 \ , \ |a_{i,j}| \le B,$$

an equation which we shall call the 'global equation'. Clearly our brute force search method would be greatly speeded up if we could find congruence conditions on the unknown exponents $a_{i,j}$. If we have a solution $a_{i,j}$ to the equation then applying the map $\phi_i$ to the equation will also give a solution to the equivalent equation in the field $\mathbb{F}_q$,

$$\phi_i(\alpha_1) \prod_{i=1}^{r_1} \phi_i(\beta_{i,1})^{a_{i,1}} + \phi_i(\alpha_2) \prod_{i=1}^{r_2} \phi_i(\beta_{i,2})^{a_{i,2}} + 1 = 0;$$

indeed we obtain $s$ such equations, one for each prime ideal. As $\mathbb{F}_q^*$ is a cyclic group, generated say by the element $g$, these last equations become

$$g^{L_{1,i}(a_{1,1},\dots,a_{r_1,1})} + g^{L_{2,i}(a_{1,2},\dots,a_{r_2,2})} + 1 = 0 \ \text{ for } \ i = 1, \dots, s,$$

where $L_{j,i}$, for $j = 1, 2$ and $i = 1, \dots, s$ are linear forms. But in these last equations the $a_{i,j}$ are only determined modulo $q - 1$. Hence if we solve the above equation in $\mathbb{F}_q^*$ we obtain congruence conditions, modulo $q - 1$, on the exponents of the possible solutions of the original global equation. Looping on all possibilities for $a_{i,j}$ modulo $q - 1$ we need only check whether this gives a solution to the $s$ equations in $\mathbb{F}_q$. Hence after $O(q^{r_1+r_2})$ steps we have a set of congruence conditions on the $a_{i,j}$ modulo $q - 1$. This will have helped us if $q$ is a lot less than $B$, as then we have eliminated all but roughly $1/q^s$ of the cases for a relatively small amount of work.

**Sieving with prime ideals of different norms.** We now suppose we choose two sets of prime ideals with each set having a common norm. The prime ideals are again chosen with respect to the constraints above. The generalization of what follows to more than two sets is obvious.

Let the common norm of the first set of $s_1$ prime ideals be $q_1 = p_1^{f_1}$ and the common norm of the second set of $s_2$ prime ideals be $q_2 = p_2^{f_2}$. Using the above method we find congruence conditions on the $a_{i,j}$ modulo $q_1 - 1$ using the first set of prime ideals. Now if we use in addition the second set we will obtain congruence conditions modulo the least common multiple, $M$, of $q_1 - 1$ and $q_2 - 1$. Hence if we set $o = M/(q_1 - 1)$ there are only $o^{r_1+r_2}$ new possibilities to check for solutions in the field $\mathbb{F}_{q_2}$ using the second set. For the first set we expect to eliminate all but $1/q_1^{s_1}$ of the possibilities; with the second set we expect to eliminate all but $1/q_2^{s_2}$ of the possibilities. Hence after sieving with these two primes we expect to be left with

$$\frac{M^{r_1+r_2}}{q_1^{s_1} q_2^{s_2}}$$

possibilities for the $a_{i,j}$ modulo $M$. This is of course rather over–optimistic as it assumes that the equations in the finite fields are in some sense 'independent', but we know that they are not as they are just reductions of the global equation. However, this rough analysis seems to be born out well in practice. We have considered roughly

$$q_1^{r_1+r_2}\left(1 + \frac{o^{r_1+r_2}}{q_1^{s_1}}\right)$$

possible solutions modulo $M$. Hence there appears to be a trade off between the number of possible solutions left, the time taken to determine these and the size of the modulus, $M$. In practice it appears that one should not worry too much about the size of the modulus. If one carries out the above procedure for enough sets then the modulus will eventually become big enough. However, the more sets (i.e. the more norms) you use the more cases will be eliminated. Hence the trick is to keep $M$ as small as possible while making the number of sets used as large as possible in the shortest amount of time. This leads us to the following strategy for choosing which norms to use.

We assume we are going to use $t$ norms, so we wish to find $t$ sets of prime ideals, the $i$th set having norm $q_i$ and $s_i$ elements. We let $M_0 = 1$ and then define $M_i$ and $o_i$ for $i = 1, \dots, t$ by

$$M_i = \text{ lcm } (M_{i-1}, q_i - 1) \ , \ o_i = M_i/M_{i-1}.$$

Then we want to minimize, for $r = r_1 + r_2$,

$$o_1^r \left(1 + \frac{o_2^r}{q_1^{s_1}}\left(1 + \frac{o_3^r}{q_2^{s_2}}(\cdots)\right)\right).$$

Hence it appears a good idea to choose $q_1$ to make $o_1^r/q_1^{s_1} = q_1^{r-s_1}$ as small as possible. It then seems best to choose $q_2, \dots, q_t$ so as to minimize $o_2, \dots, o_t$. After the $t$th set of prime ideals has been used we have determined the $a_{i,j}$ modulo $M_t$ and then hopefully this is enough, when combined with the upper bound on $A$ of the previous section, to find all solutions to the $S$-unit equation we are trying to solve.

There are further optimizations which often helps in practice. In many examples which arise out of other problems (such as the type considered in the next chapter) one has various additional pieces of information. For example, by using some automorphism of the underlying problem under consideration one is often able to deduce a set of $m$ linear equations linking the $a_{i,j}$. So there are two matrices $I \in M_{m\times r_1}(\mathbb{Z})$ and $J \in M_{m\times r_2}(\mathbb{Z})$ both having rank $m$ such that

$$I\vec{a_1} = J\vec{a_2}$$

where $\vec{a_1} = (a_{1,1}, \ldots, a_{r_1,1})^t$ and $\vec{a_2} = (a_{1,2}, \ldots, a_{r_2,2})^t$. Constraints such as these can be easily built into the above sieving method as they give rise to constraints modulo $M_i$, for $i = 1, \ldots, t$.

**Parallelization.** The above sieving algorithm works surprisingly well. It also lends itself to be implemented in parallel in a *master/slave* paradigm. The *master* process performs the sieve for the first few norms and then farms out the results to the *slaves* to sieve with the remaining norms. The slaves when they have finished pass their results back to the *master* process (or to a separate *harvester* process) which saves the results to a file. The division of the $t$ norms between the master and slave processes depends on the particular application. We do not of course alter the order in which the norms are used, only how many of them are done by the *slaves* and how many by the *master*.

Careful track, however, has to be kept of the information passed to each *slave*. Unlike the case of relation collection algorithms, for example factoring algorithms, it is crucial that each *slave* finishes the task set and passes its results back for saving. If we did not keep track of this then if a *slave* crashed for some reason we may miss some solution to our $S$-unit equation.

**Example.** We end this section with an example where we can show that there are no exceptional units in a field just by performing a sieve with prime ideals lying above three different rational primes. The field we shall consider is $K = \mathbb{Q}(\theta)$ where $\theta^6 + 2 = 0$. This field has a power integral basis in $\theta$, so its ring of integers is given by $\mathbb{Z}[\theta]$. The field has unit rank two and we can take as a pair of fundamental units

$$\eta_1 = \theta^2 + 1 \ , \ \eta_2 = \theta^5 + \theta^4 - \theta^2 - \theta - 1.$$

Hence the unit group of $K$ is given by $\mathcal{O}_K^* = \langle -1 \rangle \times \langle \eta_1 \rangle \times \langle \eta_2 \rangle$. We wish to solve the equation

$$\tau_1 + \tau_2 + 1 = 0$$

where $\tau_i \in \mathcal{O}_K^*$. We first look at how the first few non-ramified rational primes decompose. This is given in the following table where $\mathfrak{p}_i$ is a prime ideal lying above $p$ of residue degree $i$.

| Prime | 5 | 7 | 11 | 13 | 17 | 19 |
|---|---|---|---|---|---|---|
| Decomposition | $\mathfrak{p}_2\mathfrak{p}_2\mathfrak{p}_2$ | $\mathfrak{p}_6$ | $\mathfrak{p}_1\mathfrak{p}_1\mathfrak{p}_2\mathfrak{p}_2$ | $\mathfrak{p}_6$ | $\mathfrak{p}_1\mathfrak{p}_1\mathfrak{p}_2\mathfrak{p}_2$ | $\mathfrak{p}_3\mathfrak{p}_3$ |

I decided to sieve with the degree one prime ideals lying above 11, then the degree one prime ideals lying above 17 and finally the degree two prime ideals lying above 5. This means that we have

$$M_1 = 10 \ , \ M_2 = 80 \ , \ M_3 = 240.$$

After sieving with all such prime ideals we wish to determine if there are any possible solutions (modulo 240) to the equation

$$(-1)^{a_0}\eta_1^{a_1}\eta_2^{a_2} + (-1)^{b_0}\eta_1^{b_1}\eta_2^{b_2} + 1 = 0,$$

where $a_0, b_0 \in \{0,1\}$ and $a_1, a_2, b_1, b_2 \in \mathbb{Z}$.

**Sieving with ideals above** 11. There are two degree one prime ideals lying above the rational prime (11) in the ring $\mathcal{O}_K$. These give rise to two maps

$$\phi_i : K^* \to K^*_{\mathfrak{p}_i}/(\mathfrak{p}_i) \cong \mathbb{F}_{11}^* = \langle 2 \rangle,$$

given by $\phi_1(\theta) = 2$ and $\phi_2(\theta) = 9$. Our units then become

$$\phi_1(\eta_1) = 5 = 2^4 \ , \ \phi_1(\eta_2) = 8 = 2^3,$$

$$\phi_2(\eta_1) = 5 = 2^4 \ , \ \phi_2(\eta_2) = 3 = 2^8.$$

Using these two maps, applied to our unit equation, leads us to deduce the following two equations in $\mathbb{F}_{11}$:

$$\begin{aligned} 2^{5a_0+4a_1+3a_2} + 2^{5b_0+4b_1+3b_2} + 1 &\equiv 0 \pmod{11}, \\ 2^{5a_0+4a_1+8a_2} + 2^{5b_0+4b_1+8b_2} + 1 &\equiv 0 \pmod{11}. \end{aligned}$$

From these two equations we can deduce a set of possible values for $(a_0, b_0) \in \{0,1\}^2$ and $(a_1, a_2, b_1, b_2) \in (\mathbb{Z}/10\mathbb{Z})^4$. A simple computer program takes under a second to check the 40000 possible values for the $a_i$ and the $b_i$. The program returns a total of 900 simultaneous solutions to the above two equations. So we have eliminated all but 1/50th of the possible solution space using just two prime ideals.

**Sieving with ideals above** 17. We now use the 900 solutions modulo 10, which we have just obtained, to deduce the possible solutions modulo $M_2 = \text{lcm}(10, 17-1) = 80$. Just as before there are two degree one prime ideals lying above the rational prime (17). We then obtain, as before, two maps $\phi_i : K^* \to \mathbb{F}_{17}^* = \langle 3 \rangle$ given by $\phi_1(\theta) = 3$ and $\phi_2(\theta) = 14$, under which our two fundamental units become

$$\phi_1(\eta_1) = 10 = 3^3 \ , \ \phi_1(\eta_2) = 5 = 2^3,$$

$$\phi_2(\eta_1) = 10 = 3^3 \ , \ \phi_2(\eta_2) = 1 = 2^0.$$

So we wish to solve the following equations in $\mathbb{F}_{17}$, given that $(a_0, a_1, a_2)$ and $(b_0, b_1, b_2)$ come from a set of 900 possible values modulo 10,

$$\begin{aligned} 3^{8a_0+3a_1+5a_2} + 3^{8b_0+3b_1+5b_2} + 1 &\equiv 0 \pmod{17}, \\ 3^{8a_0+3a_1} + 3^{8b_0+3b_1} + 1 &\equiv 0 \pmod{17}. \end{aligned}$$

We find that there are 12825 possible solutions modulo 80. This is out of a possible total of $2^2 \cdot 80^4 = 163840000$. However, we have only tested $40000 + 900 \cdot 8^4 = 3726400$ of these. So we have reduced the total amount of work needed to determine the 12825 solutions by a factor of around 50. To actually determine the 12825 solutions took under a minute.

**Sieving with ideals above** 5. We now use the 12825 solutions modulo 80 to determine solutions modulo $M_3 = \text{lcm}(80, 25-1) = 240$. This is done by using the three degree two prime ideals lying above the rational prime (5). We now have three maps $\phi_i$ from $K^*$ to the finite field $\mathbb{F}_{25}$. We fix the following representation of the finite field

$$\mathbb{F}_{25} \cong \mathbb{F}_5[\psi]/(\psi^2+3).$$

Our three maps $\phi_i$ are then determined by

$$\phi_1(\theta) = 1 + 3\psi,\ \phi_2(\theta) = 4 + 3\psi,\ \phi_3(\theta) = \psi.$$

Our two fundamental units then become

$$\phi_1(\eta_1) = \psi,\ \phi_2(\eta_1) = 4\psi,\ \phi_3(\eta_1) = 3,$$

$$\phi_1(\eta_2) = 3 + \psi,\ \phi_2(\eta_2) = 3 + 2\psi,\ \phi_3(\eta_2) = 1 + 3\psi.$$

We then have to determine which of our 12825 solutions modulo 80 can possibly be solutions modulo 240. This means that for each solution modulo 80 we need to check another $3^4 = 81$ solutions as to whether they are solutions modulo 240. This is done by checking whether they satisfy the three images of the unit equation in $\mathbb{F}_{25}$. After around 25 minutes of computing time we find that there are no such solutions. Hence the unit equation can have no global solutions and the ring of integers has no exceptional units.

## IX.3. An $S$-unit equation in a cyclic quintic field

We shall now give a complete example of the method described above. As an example we shall take as $K$ the totally real cyclic quintic field generated by $\theta$, where $\theta$ satisfies

$$\theta^5 + \theta^4 - 4\theta^3 - 3\theta^2 + 3\theta + 1.$$

The ring of integers, $\mathcal{O}_K$, in this field has a power integral basis in $\theta$ and the unit group is given by

$$\mathcal{O}_K^* = \langle -1 \rangle \times \langle \eta_1 \rangle \times \langle \eta_2 \rangle \times \langle \eta_3 \rangle \times \langle \eta_4 \rangle,$$

where $\eta_i$ are four fundamental units which we can take to be

$$\eta_1 = \theta^2 + \theta - 1, \qquad \eta_2 = \theta^4 + \theta^3 - 3\theta^2 - 3\theta,$$
$$\eta_3 = \theta^3 + \theta^2 - 2\theta - 1, \qquad \eta_4 = \theta^4 - 3\theta^2 + 1.$$

We shall denote by $\mathfrak{p}_1, \dots, \mathfrak{p}_5$ the five valuations corresponding to the infinite places of $K$. The field $K$ has class number one and a generator for the ramified prime ideal, $\mathfrak{p}_6$, lying above 11 can be taken to be

$$\pi = -1 + \theta + 3\theta^2 - \theta^3 - \theta^4.$$

For later use we first compute the heights of all the relevant numbers. Every element $\alpha \in \{\eta_1, \dots, \eta_4, \pi\}$ is an algebraic integer with minimal polynomial of degree five; hence to compute their heights we need only compute

$$h(\alpha) = \frac{1}{5}\sum_{i=1}^{5} \max(0, \log|\alpha^{(i)}|).$$

We then compute

$$h(\eta_1) = h(\eta_2) = h(\eta_4) = 0.2884, \ h(\eta_3) = 0.3183, \ h(\pi) = 0.5117.$$

We will try and solve the equation

$$\tau_1 + \tau_2 + 1 = 0$$

where the $\tau_i$ lie in the multiplicative group generated by $\mathcal{O}_K^*$ and $\pi$. So we can write

$$\tau_1 = \pm\eta_1^{a_1}\eta_2^{a_2}\eta_3^{a_3}\eta_4^{a_4}\pi^{a_5}, \ \tau_2 = \pm\eta_1^{b_1}\eta_2^{b_2}\eta_3^{b_3}\eta_4^{b_4}\pi^{b_5}.$$

We shall try to find a small upper bound on $A = \max(|a_i|, |b_i|)$. Applying the method above we see that the first thing we need to do is to compute the constant $c_1$. We compute

$$C' = \begin{pmatrix} -0.2698 & 0.1853 & -0.6048 & 1.2566 & 0.4913 \\ -0.5203 & -0.6518 & 0.08448 & 0.1853 & 0.1853 \\ 0.1853 & -0.5203 & -0.9868 & -0.2698 & 0.0248 \\ -0.6518 & 1.2566 & 0.3350 & -0.5203 & -0.1605 \\ 0 & 0 & 0 & 0 & -2.3979 \end{pmatrix}.$$

Then we determine that we can take $c_1 = 2.9242$. Given that $t_1 = t_2 = 5$ we choose $c_3 = 0.06772$. By symmetry there is no loss of generality in assuming that $A = \max(|a_i|)$, so we take $b = 1$ and then define $h$, as above, by

$$|\tau_1|_{\mathfrak{p}_h} = \min_{1 \le i \le 6} |\tau_1|_{\mathfrak{p}_i}.$$

By Lemma IX.1 we have

$$|\tau_1|_{\mathfrak{p}_h} \le e^{-c_3 A}.$$

We now split into two cases depending on whether $\mathfrak{p}_h$ corresponds to a real place or to the finite place above the prime 11.

**$\mathfrak{p}_h$ is an infinite real place.** We have $c_4 = 1$ and so we can take $c_5 = 10.237$ and so if we set

$$\Delta_2 = b_1 \log |\eta_1^{(h)}| + \cdots + b_4 \log |\eta_4^{(h)}| + b_5 \log |\pi^h|$$

then we obtain

$$|\Delta_2| \leq c_6 e^{-c_7 A}$$

where $c_6 = 2$ and $c_7 = 0.0677$.

To bound $A$ all we need to is find a lower bound on the linear form in logarithms above. This as usual is accomplished using the result of Baker and Wüstholz, Appendix A.1. The modified heights of our elements come out to be

$$h_m(\eta_1), h_m(\eta_2), h_m(\eta_3), h_m(\eta_4) \leq 0.4796, \quad h_m(\pi) \leq 0.5117,$$

for whichever value of $h$ we take in $\{1, 2, 3, 4, 5\}$. So we obtain either $(b_1, b_2, b_3, b_4, b_5) = 0$ or

$$-c_8 \log A < \log |\Delta_2| \leq \log(c_6) - c_7 A,$$

where $c_8 = 8.4 \cdot 10^{21}$. This leads to the an upper bound of $10^{26}$ on $A$. We clearly need to reduce this upper bound. We only summarize the details here as the computations involved use LLL–reductions of $5 \times 5$ matrices with coefficients of up to 140 decimal digits.

We first apply our reduction method using five matrices and multiplying the logarithms by $10^{140}$ before rounding (there are five matrices as there is one for each embedding into the real numbers). In a matter of seconds we manage to reduce our bound to 3836.

We then apply the whole method again but this time multiply the logarithms by $10^{24}$ before rounding. Now we find we have reduced our bound to 704. A third and fourth application reduces our bound on $A$ to 604. A fifth reduction is not successful so we have a bound of 604 on $A$ in the case where $\mathfrak{p}_h$ corresponds to a real infinite place.

**$\mathfrak{p}_h$ is the finite place above 11.** Again we have $c_4 = 1$ and so $c_{10} = 0$, which means that as long as $A \neq 0$ we have

$$\text{ord}_{\mathfrak{p}_h}(\tau_1) = a_5 > 0 \text{ and } \text{ord}_{\mathfrak{p}_h}(\tau_2) = b_5 = 0.$$

This is because

$$\text{ord}_{\mathfrak{p}_h}(\tau_1) \geq 5c_9 A$$

where $c_9 = 0.00565$. We apply Yu's theorem to find $c_{11} = 3 \cdot 10^{24}$ and $c_{12} = 2 \cdot 10^{25}$ such that

$$5c_9 A \leq \text{ord}_{\mathfrak{p}_h}(\tau_1) = \text{ord}_{\mathfrak{p}_h}\left(\prod_{i=1}^{4} \eta_i^{b_i} - 1\right) \leq c_{11} \log A + c_{12}.$$

Hence we deduce that $A$ must be bounded from above by $2 \cdot 10^{29}$. Again we need to reduce this bound; this will, however, require the 11-adic logarithms of the $\eta_i$. Now if $A \geq 1/c_9 = 177.066$ then

$$\mathrm{ord}_{11}(\tau_1) = \mathrm{ord}_{11}(b_1 \log_{11} \eta_1 + \cdots + b_4 \log_{11} \eta_4) > c_9 A.$$

This last inequality is of the type we have met before in relation to linear forms in $p$-adic logarithms. Clearly the first thing that is required is to compute the 11-adic logarithms. Applying our algorithm from an earlier chapter we find

$$\begin{aligned}
\log_{11}(\eta_1) = \frac{1}{55}\log_{11}(\eta_1^{55}) &= 4\theta^4 + 7\theta^3 + 8\theta^2 + 10\theta + 4 \\
&\quad +11(6\theta^4 + 8\theta^3 + 8\theta^2 + 10\theta + 3) + O(11^2), \\
\log_{11}(\eta_2) = \frac{1}{110}\log_{11}(\eta_2^{110}) &= 9\theta^4 + 2\theta^3 + 3\theta^2 + 9\theta + 8 \\
&\quad +11(4\theta^4 + 7\theta^3 + 5\theta^2 + 8\theta + 4) + O(11^2), \\
\log_{11}(\eta_3) = \frac{1}{110}\log_{11}(\eta_3^{110}) &= 5\theta^3 + 2\theta^2 + 7 \\
&\quad +11(10\theta^4 + 7\theta^3 + 4\theta^2 + 6\theta + 4) + O(11^2), \\
\log_{11}(\eta_4) = \frac{1}{55}\log_{11}(\eta_4^{55}) &= 6\theta^4 + 8\theta^3 + 2\theta + 1 \\
&\quad +11(3\theta^1 1 + 8\theta^3 + 3\theta^2 + 9\theta + 1) + O(11^2).
\end{aligned}$$

For what follows we need to compute the values of these 11-adic logarithms to around sixty 11-adic places of accuracy. This, however, does not take too long on a computer. We then apply our $p$-adic reduction algorithm. Rounding the logarithms to 55 digits accuracy at first reduces our bound to 2018. Repeating the process using 7 digits accuracy we obtain a new bound of 318. Finally using an accuracy of 5 digits we manage to reduce the bound to 247.

**The sieving step.** So to sum up the two cases we have

$$\tau_1 + \tau_2 + 1 = 0$$

where

$$\tau_1 = \pm\eta_1^{a_1}\eta_2^{a_2}\eta_3^{a_3}\eta_4^{a_4}\pi^{a_5}, \quad \tau_2 = \pm\eta_1^{b_1}\eta_2^{b_2}\eta_3^{b_3}\eta_4^{b_4}\pi^{b_5}.$$

and $A = \max(|a_i|, |b_i|) \leq 604$. Alas, with the computing power available (and in a reasonable amount of time) it was impossible to find all the solutions to this unit equation. This is not because the bound of 604 is too high but because the number of exponents, i.e. 10, is too large without any extra structure being present.

In any case we were able to solve the problem when $a_5 = b_5 = 0$, i.e. $\tau_1, \tau_2 \in \mathcal{O}_K^*$. In other words we were able to find all exceptional units in the field $K$. We could repeat the above analysis again to derive a new upper bound on $A$ which would be smaller than our upper bound of 604 in this new situation, a process which we leave as an exercise. We used the above bound of 604 to find the exceptional units.

To apply the sieving method we used the following finite fields:

$$\mathbb{F}_{23}, \mathbb{F}_{67}, \mathbb{F}_{199}, \mathbb{F}_{397}, \mathbb{F}_{89}, \mathbb{F}_{109}, \mathbb{F}_{353}, \mathbb{F}_{331}, \mathbb{F}_{241}, \mathbb{F}_{43}$$

which were used in the order given above. Notice then that the values of $o_i$ with this ordering are

$$22, 3, 3, 2, 2, 3, 4, 5, 1, 7,$$

which explains why we choose the above ordering as we want to keep the $o_i$ as small as possible. After a couple of hours of computing time the program returns a list of 770 possible exponent vectors for the exceptional units. These are only possible solutions, i.e. ones which pass the sieve for all the above finite fields.

The exponent vectors then need to be checked as to whether they actually give rise to exceptional units. This is a trivial task which can be accomplished in any computer algebra system. In the end we found a total of 570 exceptional units in the field $K$. These solutions all had exponents which satisfied $|a_i|, |b_i| \leq 7$.

## IX.4. Integral points on elliptic curves (II)

We now present our second algorithm to find integral points on elliptic curves. The method of this section dates back to Siegel and, of the three methods we present in this book, the following is probably the most inefficient. As before we wish to find all integral solutions to the equation

$$Y^2 = F(X)$$

where $F(X)$ is a monic cubic polynomial with integral coefficients and non-zero discriminant. We shall denote the roots of $F(X)$ by $\theta_1, \theta_2, \theta_3$.

As before we can, after a little bit of algebra, determine a finite number of possibilities for the $\alpha_i$ in the following equation:

$$X - \theta_i = \alpha_i \beta_i^2 \text{ for } i = 1, 2, 3.$$

For each of these finite number of possibilities for the $\alpha_i$ we look at the extension of $K = \mathbb{Q}(\theta_1, \theta_2, \theta_3)$ given by $L = K(\sqrt{\alpha_1}, \sqrt{\alpha_2}, \sqrt{\alpha_3})$. This is a field extension of $K$ unramified away from the prime ideals of $K$ which divide the $\alpha_i$. Writing $\gamma_i = \sqrt{\alpha_i}$ we have

$$X - \theta_i = (\gamma_i \beta_i)^2$$

and hence

$$\theta_j - \theta_i = (\gamma_i \beta_i)^2 - (\gamma_j \beta_j)^2 = (\gamma_i \beta_i - \gamma_j \beta_j)(\gamma_i \beta_i + \gamma_j \beta_j).$$

We let $S$ denote the set of places of $L$ which are either infinite or lie above primes of $\mathbb{Q}$ which divide the discriminant of $F(X)$ or which lie above the prime 2. By definition $\theta_j - \theta_i$ is an element of $\mathcal{O}_S^*$ and hence so is each factor on the right hand side of the previous equation. Hence if we set

$$\tau_1^{(\pm)} = \frac{\gamma_1\beta_1 \pm \gamma_2\beta_2}{\gamma_1\beta_1 - \gamma_3\beta_3} \text{ and } \tau_2^{(\pm)} = \frac{\gamma_2\beta_2 \pm \gamma_3\beta_3}{\gamma_1\beta_1 - \gamma_3\beta_3}$$

then we have that $\tau_1, \tau_2 \in \mathcal{O}_S^*$. But $\tau_1$ and $\tau_2$ are related by the equation

$$-\tau_1^{(\pm)} \pm \tau_2^{(\pm)} + 1 = 0,$$

which is a two term $S$-unit equation in $L$. Hence by the above algorithm we can determine all solutions for $\tau_1$ and $\tau_2$. Therefore we can compute all possible values for

$$\begin{aligned} \sigma = \sqrt{\frac{\theta_2 - \theta_1}{\tau_1^{(+)}\tau_1^{(-)}}} &= \sqrt{\frac{(\theta_2 - \theta_1)(\gamma_1\beta_1 - \gamma_3\beta_3)^2}{(\gamma_1\beta_1 + \gamma_2\beta_2)(\gamma_1\beta_1 - \gamma_2\beta_2)}} \\ &= \sqrt{(\gamma_1\beta_1 - \gamma_3\beta_3)^2} = \pm(\gamma_1\beta_1 - \gamma_3\beta_3). \end{aligned}$$

So we can compute

$$\begin{aligned} \theta_1 + \frac{1}{4}\left(\pm\sigma + \frac{\theta_3 - \theta_1}{\sigma}\right)^2 &= \theta_1 + \frac{1}{4}\left(\frac{\sigma^2 \pm \sigma(\gamma_1\beta_1 + \gamma_3\beta_3)}{\sigma}\right)^2 \\ &= \theta_1 + \frac{1}{4}(\pm\sigma + \gamma_1\beta_1 + \gamma_3\beta_3)^2 \\ &= \theta_1 + (\gamma_1\beta_1)^2 = X \end{aligned}$$

which will give us all our integral points $(X, Y)$. However, one should note that the above method is rather silly in practice as we have to go to all the expense of solving an $S$-unit equation in the number field $L$. This could in general be a degree 24 extension of the rational numbers which is rather large to work with using current technology, given we already have given an easier method in Chapter VII and we shall present an even better method in Chapter XIII. We shall therefore not present an example of this method at all.

## IX.5. Other applications

The two term $S$-unit equation which we considered in this chapter has a wide range of applications to various areas of number theory; As we pointed out before they occur in the solution of Thue and Thue–Mahler equations. In the next couple of chapters we shall see how they occur in solving various other types of diophantine equations.

They also occur when one tries to enumerate binary forms of given discriminant. The case of quadratic and cubic forms are classical and can be traced back to Lagrange [**109**] and Hermite [**98**]. That there are only finitely many binary forms of given degree and discriminant follows from a theorem of Birch and Merriman [**14**]. Birch and Merriman's proof, however, is ineffective even though it makes use of the fact that an $S$-unit equation has only finitely many solutions.

An effective proof of the finiteness of the number of binary forms of given degree and discriminant was given by Evertse and Győry [**52**]. Their proof not only uses the finiteness of the number of solutions of $S$-unit equations but also a more complicated analysis than Birch and Merriman in order to make the proof effective. Even though effective, a lot of work has to be done to apply Evertse and Győry's method in practice. However, there are various optimizations which can be performed. This allows in some cases explicit numerical calculations to be performed, see [**182**].

One application of the determination of binary forms with given discriminant is to find hyperelliptic curves of genus greater than one with bad reduction at only a finite set of primes. That there are only finitely many follows from the theorem of Faltings [**55**]; however, the proof is ineffective. Various authors [**126**], [**198**], [**129**] had tried to determine all curves of genus 2 with good reduction away from 2. However, it was not until a practical version of Evertse and Győry's algorithm was implemented that the list of all such curves could be completed, see again [**182**].

One of the most interesting types of two term unit equations is of the form

$$\tau_1 + \tau_2 + 1 = 0 \ , \ \tau_i \in \mathcal{O}_K^*.$$

Solutions to this equation are called exceptional units. There is a link between the number of exceptional units a number field has and whether it is a euclidean field. This link is due to Lenstra [**118**]. Because of this link a lot of computational work has been done on exceptional units. The interested reader should consult [**140**], [**50**], [**121**] and [**141**]. There is even a link between exceptional units, Lenstra's work and the dynamics of iterated polynomial mappings [**220**].

There has been some work on determining exceptional units in parameterized families of number fields. For a survey of techniques in this area see [**141**]. For an example see [**142**], where the family of fields, $\mathbb{Q}(\theta_a)$, are considered, for $a \in \mathbb{Z}_{>0}$, with

$$\theta_a^4 + a\theta_a^3 + \theta_a^2 + a\theta_a - 1 = 0.$$

In [**150**] a family of quartic fields are considered which depend on two, rather than one, parameter.

## IX.6. Exercise

1). Deduce a smaller upper bound than 604 for the exponents, in our $S$-unit equation in the cyclic quintic field considered above, for the case when $a_5 = b_5 = 0$.

CHAPTER X

# Triangularly connected decomposable form equations

In this chapter we apply the method to solve $S$-unit equations, developed in the previous chapter, to solve a class of diophantine equations called triangularly connected decomposable form (TCDF) equations. These were first studied by Győry, see [**89**], [**93**], [**92**] and [**91**], who gave effective upper bounds on their solutions. A practical algorithm was given in [**180**], and it is this method which we shall explain here. We shall see that both Thue and Thue–Mahler equations are examples of TCDF equations. Other equations also fall into this category, for instance discriminant and index form equations. While we shall develop the following for forms with integer coefficients and variables, almost all of what we shall say goes over verbatim to when the coefficients of the form and the variables lie in some ring of integers of a number field. For instance this allows one to solve Thue and Thue–Mahler equations defined over a ring of integers [**183**]. However, there is often a much better method, see for instance [**211**].

## X.1. Triangularly connected linear forms

We first consider what it means for a set of linear forms to be called triangularly connected. We shall also show how one can easily determine whether or not a set of linear forms is triangularly connected or not.

Let $\mathcal{L}$ denote a set of $m$ linear forms in $v$ variables with coefficients in the ring of integers of some number field $K$. We shall suppose that $m \geq 3$ and that $[K : \mathbb{Q}] = n$. Each linear form $L_j(\vec{x}) \in \mathcal{L}$ we shall write as

$$L_j(\vec{x}) = \sum_{i=1}^{v} \ell_{i,j} x_i \text{ where } \ell_{i,j} \in \mathcal{O}_K$$

Győry [**92**] called such a set triangularly connected if for all $i, j \in \{1, \ldots, m\}$, such that $i \neq j$, there is a sequence of linear forms in $\mathcal{L}$, say

$$L_i = L_{i_1}, L_{i_2}, \ldots, L_{i_w} = L_j$$

such that for each $u \in \{1, \ldots, w-1\}$ there exists non-zero $\alpha_k \in \mathcal{O}_K$ for $k = 1, 2, 3$ and a linear form $L_{i_{u,u+1}} \in \mathcal{L}$ such that

$$\alpha_1 L_{i_u}(\vec{x}) + \alpha_2 L_{i_{u+1}}(\vec{x}) + \alpha_3 L_{i_{u,u+1}}(\vec{x}) \equiv 0.$$

This may look rather daunting at first but it is actually quite natural when one considers why we call such a set triangularly connected. Let $\mathbb{G}_{\mathcal{L}}$

be the hyper-graph which has as vertices all the $L_i \in \mathcal{L}$. For three such vertices $L_1, L_2, L_3$ we connect the vertices with a hyper-edge (or one might say 'triangle') if and only if there exists $\alpha_k \in \mathcal{O}_K$, for $k = 1, 2, 3$, such that

$$\alpha_1 L_1(\vec{x}) + \alpha_2 L_2(\vec{x}) + \alpha_3 L_3(\vec{x}) \equiv 0,$$

with $\alpha_1\alpha_2\alpha_3 \neq 0$. So the hyper-graph contains only 'triangles' as hyper-edges, and it is a connected hyper-graph if and only if the set of linear forms $\mathcal{L}$ is itself triangularly connected by the definition above.

We now need to give an algorithm to determine whether a set of linear forms is triangularly connected. We do this by determining all the hyper-edges in the hyper-graph, $\mathbb{G}_{\mathcal{L}}$. The following method will accomplish this. For each of the ${}^mC_3$ possible choices for $L_1, L_2, L_3 \in \mathcal{L}$ we need to determine whether there is a solution to

$$\alpha_1 L_1(\vec{x}) + \alpha_2 L_2(\vec{x}) + \alpha_3 L_3(\vec{x}) \equiv 0 \tag{X.1}$$

with $\alpha_1\alpha_2\alpha_3 \neq 0$ and $\alpha_k \in \mathcal{O}_K$. Suppose that $\mathcal{O}_K$ has an integral basis given by $\omega_1, \ldots, \omega_n$, then it is easy to compute the 'multiplication table', $\Gamma \in \mathbb{Z}^{n^3}$, of such an integral basis from

$$\omega_a\omega_b = \sum_{c=1}^{n} \Gamma(c, a, b)\omega_c.$$

We write

$$L_i(\vec{x}) = \sum_{j=1}^{n}\left(\sum_{k=1}^{v} l_{i,j,k}x_k\right)\omega_j \text{ for } i = 1, 2, 3$$

where $l_{i,j,k} \in \mathbb{Z}$. We also write out the unknowns as

$$\alpha_i = \sum_{a=1}^{n} y_{i,a}\omega_a \text{ for } i = 1, 2, 3$$

where the $y_{i,a}$ are integers to be found. Equation (X.1) then implies that we have

$$\sum_{a=1}^{n}\omega_a\left(\sum_{b=1}^{v}x_b\left(\sum_{c,d=1}^{n}\Gamma(a, c, d)\left(l_{1,d,b}y_{1,d} + l_{2,d,b}y_{2,d} + l_{3,d,b}y_{3,d}\right)\right)\right) = 0.$$

This must be satisfied for all possible values of $\vec{x}$ hence we can equate coefficients of $\omega_a$ and $x_b$ in the above equation to obtain $nv$ linear equations in the $3n$ unknowns:

$$\sum_{c,d=1}^{n}\Gamma(a, c, d)\left(l_{1,d,b}y_{1,d} + l_{2,d,b}y_{2,d} + l_{3,d,b}y_{3,d}\right) = 0 \text{ for } 1 \leq a \leq n,\ 1 \leq b \leq v.$$

These equations can then be solved by standard linear algebra over the ring $\mathbb{Z}$. If there are no solutions then we know that (X.1) has no solution. If, however, there are solutions then $L_1, L_2$ and $L_3$ are connected by a hyper-edge (triangle) in the hyper-graph, and we also have determined explicit values of $\alpha_1, \alpha_2$ and $\alpha_3$ to use in future calculations.

## X.2. TCDF equations

Having defined what one means by a set of linear forms to be triangularly connected we now need to decide what it means for a decomposable form to be triangularly connected.

A form $F(\vec{x})$ in $v$ variables of degree $m$ is called decomposable if it can be factored over the algebraic closure into a product of linear forms, i.e. it can be decomposed into linear forms:

$$F(\vec{x}) = a_0 L'_1(\vec{x}) \cdots L'_m(\vec{x})$$

where $a_0 = F(1, 0, \ldots, 0) \in \mathbb{Z}$ and $L'_i(\vec{x})$ are linear forms with coefficients in a Galois extension, $K$, of $\mathbb{Q}$ of degree $n$. We can write

$$L'_j(\vec{x}) = x_1 + \ell'_{2,j} x_2 + \cdots + \ell'_{v,j} x_v$$

where $\ell'_{i,j} \in K$. If we set $G = Gal(K/\mathbb{Q})$ then the set $\mathcal{L}' = \{L'_1, \ldots, L'_m\}$ can be chosen to be stable as a set with respect to $G$. In other words if $L'_i$ is a linear form in $\mathcal{L}'$ and $\sigma \in G$ then $\sigma(L'_i) \in \mathcal{L}'$. We note that the factorization of a decomposable form into its linear decomposition can be achieved using the method in [**148**]. However, in the examples we shall consider, the factorization into linear factors is trivial.

If we let $\ell_{i,j} = a_0 \ell'_{i,j}$ for $i \geq 2$ and $\ell_{1,j} = a_0$ then $\ell_{i,j} \in \mathcal{O}_K$ for all $i, j$. Upon setting

$$L_j(\vec{x}) = a_0 L'_j(\vec{x}) = \sum_{i=1}^{v} \ell_{j,i} x_i$$

the set $\mathcal{L} = \{L_1(\vec{x}), \ldots, L_m(\vec{x})\}$ is stable with respect to $G$. The form $F(\vec{x})$ will be called triangularly connected if the set $\mathcal{L}$ is a triangularly connected set.

LEMMA X.1. *Every binary form of degree greater than two with non-zero discriminant is triangularly connected.*

PROOF. Exercise. □

From now on we shall suppose that $F(\vec{x})$ is a decomposable form which is triangularly connected, i.e. $F(\vec{x})$ is a TCDF. We wish to find all solutions, $\vec{x} \in \mathbb{Z}^v$, to the equation

$$F(\vec{x}) = A p_1^{z_1} \cdots p_t^{z_t} \tag{X.2}$$

subject to $\gcd(\vec{x}) = 1$ and $z_i \in \mathbb{N}$. We assume that $A \in \mathbb{Z}$ is given, as are the prime numbers $p_i$. The above TCDF equation is said to be of Thue type if $t = 0$ and of Mahler type otherwise (the reason is of course historically obvious). Let $\mathfrak{p}_1, \ldots, \mathfrak{p}_s$ denote the prime ideals in $K$ which lie above the prime numbers $p_1, \ldots, p_t$. We define $\pi_i \in \mathcal{O}_K$ to be a generator of the ideal $\mathfrak{p}_i^{h_K}$ where $h_K$ is the class number of $K$. As usual we let $e_i$ denote the ramification index of $\mathfrak{p}_i$ over $\mathbb{Q}$.

Clearly solving equation (X.2) is equivalent to solving the equation

$$f(\vec{x}) = Cp_1^{z_1} \cdots p_t^{z_t} \tag{X.3}$$

where $f(\vec{x})$ is the TCDF

$$f(\vec{x}) = a_0^{m-1} F(\vec{x}) = \prod_{j=1}^{m} L_j(\vec{x})$$

and $C = Aa_0^{m-1}$. We shall need to make the following assumption on the form $f(\vec{x})$,

**Assumption.** There is no $\vec{x} \in \mathbb{Z}^v$, with $\vec{x} \neq \vec{0}$, such that $L_j(\vec{x}) = 0$ for all $j$.

This means that the matrix

$$\begin{pmatrix} \ell_{1,1} & \cdots & \ell_{v,1} \\ \vdots & & \vdots \\ \ell_{1,m} & \cdots & \ell_{v,m} \end{pmatrix}$$

has column rank $v$ over $\mathbb{Q}$.

LEMMA X.2. *Under the above assumption we have $v \leq mn$.*

PROOF. Exercise. □

## X.3. Solving TCDF equations

In this section we consider the solution of TCDF equations. Firstly we reduce the problem to determining the solutions of a large finite set of $S$-unit equations. Then we describe how to reduce the total number of $S$-unit equations that need to be solved to something more manageable. Finally we describe techniques to make the $S$-unit equations easier to solve. These last two techniques make use of the Galois group $G$, introduced earlier.

**X.3.1. Reduction to a finite set of $S$-unit equations.** Owing to the assumption made above we can find an index set $I = \{i_1, \ldots, i_v\}$ with $1 \leq i_k \leq m$ and an index set $J = \{j_1, \ldots, j_v\}$ with $1 \leq j_k \leq n$ such that the following matrix is invertible:

$$A_{I,J} = \begin{pmatrix} \ell_{1,i_1}^{(j_1)} & \cdots & \ell_{v,i_1}^{(j_1)} \\ \vdots & & \vdots \\ \ell_{1,i_v}^{(j_v)} & \cdots & \ell_{v,i_v}^{(j_v)} \end{pmatrix}.$$

Clearly we then have that $\det(A_{I,J}) \in \mathcal{O}_K$ and we can set

$$c_1 = |N_{K/\mathbb{Q}}(\det(A_{I,J}))|.$$

We shall require the following lemma:

LEMMA X.3. *Suppose that there exist $a \in \mathbb{Z}$ and $\xi_j \in \mathcal{O}_K$ such that for some value $\vec{x} \in \mathbb{Z}^v$ we have*

$$L_j(\vec{x}) = a\xi_j$$

*then* $|a|^n \leq c_1$.

PROOF. With the choice of $I$ and $J$ in the definition of $c_1$ above we have

$$A_{I,J}\vec{x} = a\vec{b}$$

where

$$\vec{b} = \begin{pmatrix} \xi_{i_1}^{(j_1)} \\ \vdots \\ \xi_{i_v}^{(j_v)} \end{pmatrix}.$$

Then by Cramer's rule we obtain

$$x_i = a\det(A_{I,J}^{(i)})/\det(A_{I,J})$$

where $A_{I,J}^{(i)}$ is the matrix obtained from $A_{I,J}$ by replacing column $i$ of $A_{I,J}$ by the vector $\vec{b}$. If we then set $\alpha_i = \det(A_{I,J}^{(i)})$ and $\alpha = \det(A_{I,J})$ then it is clear that $\alpha_i, \alpha \in \mathcal{O}_K$. The equation $x_i = a\alpha_i/\alpha$ implies that we have the following ideal equation:

$$(a) \cdot (\alpha_1, \dots, \alpha_v) = (x_1, \dots, x_v) \cdot (\alpha) = (\alpha),$$

and so $|a|^n = |N_{K/\mathbb{Q}}(a)| \leq |N_{K/\mathbb{Q}}(\alpha)| = c_1$. □

We use this result to prove the following result

LEMMA X.4. *Suppose $\vec{x} \in \mathbb{Z}^v$ is a solution to our TCDF equation (X.3) and we have*

$$L_j(\vec{x}) = \sigma\delta_j$$

*where $\delta_j \in \mathcal{O}_K$ is given and*

$$\sigma = \epsilon\pi_1^{a_1}\cdots\pi_s^{a_s},$$

*with $\epsilon \in \mathcal{O}_K^*$ and $a_i \in \mathbb{N}$ unknown. In such a situation the $a_i$ are bounded by*

$$0 \leq a_i \leq e_i\left(1 + c_2(k)\right)/h_K + c_3(k),$$

*where $p_k$ is the rational prime lying below $\mathfrak{p}_i$ and*

$$\begin{aligned} c_2(k) &= (\log c_1)/(p_k n) \\ c_3(k) &= \frac{1}{mh_K}\max_{\mathfrak{q}|p_k}\left(\mathrm{ord}_{\mathfrak{q}}\left(\prod_{j=1}^m \delta_j\right)\right). \end{aligned}$$

PROOF. Let $\mathfrak{p}$ denote a prime ideal of $K$ lying above $p_k$. We set $g_k = \mathrm{ord}_{p_k}(C) + z_k$ and we note that the ramification index of $\mathfrak{p}$ only depends on $p_k$ and not $\mathfrak{p}$ as $K$ is a Galois field. We shall therefore denote this ramification index by $e_k$. We shall denote by $d_k$ the greatest rational integer such that the inequality

$$g_k e_k - \operatorname{ord}_{\mathfrak{p}} \left( \prod_{j=1}^{m} \delta_j \right) \geq m d_k e_k \tag{X.4}$$

holds for every prime ideal $\mathfrak{p}$ lying above $p_k$. By this definition of $d_k$ there exists a prime ideal $\mathfrak{q}$ which divides $p_k$ such that

$$m(d_k + 1)e_k > g_k e_k - \operatorname{ord}_{\mathfrak{q}} \left( \prod_{j=1}^{m} \delta_j \right). \tag{X.5}$$

We also have the equality, for the prime ideal $\mathfrak{p}$ dividing $\pi_i$ and $p_k$,

$$a_i m h_K + \operatorname{ord}_{\mathfrak{p}} \left( \prod_{j=1}^{m} \delta_j \right) = g_k e_k \tag{X.6}$$

Hence, by (X.4) and (X.5),

$$\begin{aligned} 0 \ &\leq \ d_k e_k / h_K \leq \frac{1}{m h_K} \left( g_k e_k - \operatorname{ord}_{\mathfrak{p}} \left( \prod_{j=1}^{m} \delta_j \right) \right), \\ &= \ a_i, \\ &\leq \ g_k e_k / m h_K < (d_k + 1) e_k / h_K + \frac{1}{m h_K} \operatorname{ord}_{\mathfrak{q}} \left( \prod_{j=1}^{m} \delta_j \right) \end{aligned}$$

We therefore need now only show that

$$d_k \leq (\log c_1)/(p_k n).$$

To accomplish this we let $a \in \mathbb{Z}$ denote a number such that $a = p_1^{d_1} \cdots p_t^{d_t}$ and choose $\xi \in K$ such that $a\xi = \pi_1^{a_1} \cdots \pi_s^{a_s}$. It follows that

$$e_k d_k + \operatorname{ord}_{\mathfrak{p}}(\xi) = \operatorname{ord}_{\mathfrak{p}}(a\xi) = h_K a_i,$$

where $\mathfrak{p}$ is the prime ideal dividing both $\pi_i$ and $p_k$. Then

$$\begin{aligned} \operatorname{ord}_{\mathfrak{p}}(\xi) \ &= \ h_K a_i - e_k d_k \\ &= \ \frac{1}{m} \left( g_k e_k - \operatorname{ord}_{\mathfrak{p}} \left( \prod_{j=1}^{m} \delta_j \right) \right) - e_k d_k \text{ by (X.6)} \\ &> \ d_k e_k - e_k d_k = 0 \text{ by (X.4)} \end{aligned}$$

Hence $\xi \in \mathcal{O}_K$, and writing $\xi_j = \epsilon \xi \delta_j \in \mathcal{O}_K$ we have

$$L_j(\vec{x}) = a\xi_j$$

We are then in the situation described by Lemma X.3, so we can deduce $|a|^n \leq c_1$ and so we obtain the required bound on $a_i$. □

Note we can deduce something even stronger which can help in practice. We find that

$$0 \le a_i \le e_k(1+d_k)/h_K + c_3(k)$$

and

$$\sum_{i=1}^{t} d_k \log p_k \le \frac{\log c_1}{n}.$$

In any case, if we can apply the last lemma, we know that we can write for all $j$, $1 \le j \le m$,

$$L_j(\vec{x}) = \epsilon\gamma_j$$

where the $\gamma_j$s are some given known quantities and $\epsilon$ is some unknown element of $\mathcal{O}_K^*$. We have then the equation

$$\epsilon^m \prod_{j=1}^{m} \gamma_j = Cp_1^{z_1}\cdots p_t^{z_t}.$$

Taking $\mathrm{ord}_{p_i}$ of both sides we have therefore determined the $z_i$. We write

$$\epsilon = \xi\eta_1^{v_1}\cdots\eta_r^{v_r},$$

for a unit of finite order $\xi$ and a set of fundamental units $\eta_1,\dots,\eta_r$. These values can then be read off the equation

$$(\xi\eta_1^{v_1}\cdots\eta_r^{v_r})^m = Cp_1^{z_1}\cdots p_t^{z_t}\prod_{j=1}^{m}\gamma_j^{-1}.$$

If no such integer values of $v_i$ are possible then this set of $\gamma_j$'s does not correspond to a solution. So finally, assuming the conditions of Lemma X.4 are satisfied, we can determine the values of $\tau_j = L_j(\vec{x})$. Our final task is to read off the values of $\vec{x}$. Let $\mathcal{O}_K$ have an integral basis $\omega_1,\dots,\omega_n$. We can then write

$$\tau_i = \sum_{j=1}^{n} t_{i,j}\omega_j \;\; \ell_{i,j} = \sum_{k=1}^{n} l_{i,j,k}\omega_k$$

where $t_{i,j}, l_{i,j,k} \in \mathbb{Z}$ are known. Hence we wish to determine the values of $x_i$ for $i = 1,\dots,v$ from the $mn$ equations given by

$$\sum_{i=1}^{v} \ell_{i,j,k}x_i = t_{j,k} \text{ for } j \text{ and } k \text{ such that } 1 \le j \le m\ ,\ 1 \le k \le n.$$

All that remains is then to show that we can get ourselves into a position to be able to apply Lemma X.4.

THEOREM X.5. *Suppose $\vec{x} \in \mathbb{Z}^v$ is a solution to our TCDF equation (X.3). We can then find sets of $\delta_j \in \mathcal{O}_K$ such that*

$$L_j(\vec{x}) = \sigma\delta_j$$

*where $\sigma$ is given by*

$$\sigma = \epsilon\pi_1^{a_1}\cdots\pi_s^{a_s},$$

*with $\epsilon \in \mathcal{O}_K^*$ and $a_i \in \mathbb{N}$ unknown.*

PROOF. We have

$$\prod_{j=1}^{m} L_j(\vec{x}) = Cp_1^{z_1}\cdots p_t^{z_t}.$$

Hence by the unique factorization of ideals there is a finite number of possibilities for ideals $\mathfrak{a}_j$ such that

$$L_j(\vec{x})\mathcal{O}_S^* = \mathfrak{a}_j\mathcal{O}_S^*.$$

Firstly we observe that if we have an equation of the form

$$\alpha_1 L_1(\vec{x}) + \alpha_2 L_2(\vec{x}) + \alpha_3 L_3(\vec{x}) = 0$$

then, by our algorithm for $S$-unit equations of the preceding chapter, we can determine $L_1(\vec{x}), L_2(\vec{x}), L_3(\vec{x})$ up to multiplication by an unknown element of $\mathcal{O}_S^*$. We shall use this observation in what follows.

As our form is triangularly connected, for any index, $j$, we care to consider we can find a sequence of the linear forms $L_j(\vec{x})$,

$$L_1(\vec{x}) = L_{i_1}(\vec{x}), \ldots, L_{i_w}(\vec{x}) = L_j(\vec{x}),$$

such that for all $u \in \{1, \ldots, w-1\}$ there exist $\alpha_{i_u}, \alpha_{i_{u+1}}, \alpha_{i_u,u+1} \in \mathcal{O}_K^*$ such that for some linear form $L_{i_u,u+1}(\vec{x}) \in \mathcal{L}$ we have the equation

$$\alpha_{i_u} L_{i_u}(\vec{x}) + \alpha_{i_{u+1}} L_{i_{u+1}}(\vec{x}) + \alpha_{i_u,u+1} L_{i_u,u+1}(\vec{x}) = 0.$$

You should think of this sequence as a set of stepping stones jumping from $L_1(\vec{x})$ to $L_j(\vec{x})$ in the graph $\mathbb{G}_{\mathcal{L}}$. We can step from one vertex to another only if both vertices are contained in a hyper-edge (triangle).

Applying our observation above and solving $w$ such $S$-unit equations we can find $\kappa_i \in \mathcal{O}_K$ such that

$$\begin{array}{ll} L_1(\vec{x}) = \sigma_1\kappa_1 & L_2(\vec{x}) = \sigma_1\kappa_2 \\ \vdots & \vdots \\ L_{w-1}(\vec{x}) = \sigma_{w-1}\kappa_{w-1,i_{w-1}} & L_w(\vec{x}) = \sigma_{w-1}\kappa_{w-1,i_w} \end{array}$$

where the $\sigma_i$ are unknown elements of $\mathcal{O}_S^*$. We can then write for all $j$, with $1 \le j \le m$,

$$\begin{aligned} L_j(\vec{x}) = L_{i_w}(\vec{x}) &= \kappa_{w-1,i_w}\sigma_{w-1} \\ &= \kappa_{w-1,i_w} L_{w-1}(\vec{x})/\kappa_{w-1,i_{w-1}} \\ &= \cdots \\ &= \sigma_1\lambda_j \end{aligned}$$

where

$$\lambda_j = \kappa_2 \prod_{u=1}^{w-1} \frac{\kappa_{u,i_{u+1}}}{\kappa_{u,i_u}}.$$

Let $\Delta$ denote the group

$$\Delta \cong \mathcal{O}_S^*/(\langle\pi_1\rangle \times \cdots \times \langle\pi_s\rangle \times \mathcal{O}_K^*).$$

Then $\Delta$ is clearly finite and computable. We can write, for some $\delta \in \Delta$,

$$\sigma_1 = \epsilon\delta\pi_1^{c_1} \cdots \pi_s^{c_s}$$

where $\epsilon \in \mathcal{O}_K^*$ and the $c_i$s are unknown integers. For $1 \leq k \leq s$ we choose $b_k$ to be the smallest integer which satisfies

$$b_k h_K \geq -\mathrm{ord}_{\mathfrak{p}_k}(\delta\lambda_i) \text{ for all } 1 \leq i \leq m. \tag{X.7}$$

Suppose that $c_k < b_k$. Then for some $j$ we have

$$\begin{aligned} -\mathrm{ord}_{\mathfrak{p}_k}(\delta\lambda_j) &= h_K c_k - \mathrm{ord}_{\mathfrak{p}_k}(\sigma_1\lambda_j) \\ &\leq h_K(b_k - 1) - \mathrm{ord}_{\mathfrak{p}_k}(\sigma_1\lambda_j) \end{aligned}$$

But by choice of $b_k$ this would imply that $\mathrm{ord}_{\mathfrak{p}_k}(\sigma_1\lambda_j) < 0$ and hence $L_j(\vec{x}) \notin \mathcal{O}_K$. Since this is clearly nonsense we must have $c_k \geq b_k$, and so $a_k = c_k - b_k \geq 0$. The result follows upon setting

$$\delta_j = \pi_1^{b_1} \cdots \pi_s^{b_s}\delta\lambda_j$$

for $1 \leq j \leq m$. The $\delta_j$ coming from a finite set as the set $\Delta$ is finite and the $b_i$ are uniquely determined by $\delta \in \Delta$, the $\lambda_i$ and inequality (X.7). □

**X.3.2. Reduction of the number of equations.** From the proof of Theorem X.5 it appears that we need to solve an awfully large number of $S$-unit equations to solve our TCDF equation. If one were to use the method of this chapter to solve Thue–Mahler equations, of degree $m$, then it appears that one would need to solve ${}^mC_3$ $S$-unit equations (solving Thue–Mahler equations using the method in this chapter could be considered silly but we shall skim over this point; as long as you get the message that the number of possible $S$-unit equations could become stratospheric).

Obviously we need a way of only needing to consider a very small number of $S$-unit equations. Firstly consider that each equation which we have to solve represents a hyper-edge in our hyper-graph $\mathbb{G}_{\mathcal{L}}$. Let the set of such hyper-edges be denoted $\mathcal{E}$. Each element of $\mathcal{E}$ represents three linear forms $L_i(\vec{x}), L_j(\vec{x}), L_k(\vec{x}) \in \mathcal{L}$ and three constants $\alpha_i, \alpha_j, \alpha_k \in \mathcal{O}_K$ such that

$$\alpha_i L_i(\vec{x}) + \alpha_j L_j(\vec{x}) + \alpha_k L_k(\vec{x}) = 0.$$

To help in our task of reducing the number of $S$-unit equations we need to consider, we bring in the Galois group of $K$ over $\mathbb{Q}$, $G = Gal(K/\mathbb{Q})$. We have already commented that $\mathcal{L}$ is stable as a set under the action of $G$. Let $\sigma$ denote an arbitrary element of $G$ and then set

$$\alpha_i' = \sigma(\alpha_i) \in \mathcal{O}_K \text{ , } L_i'(\vec{x}) = \sigma(L_i(\vec{x}) \in \mathcal{L}.$$

But then we have the equation

$$\alpha_i' L_i'(\vec{x}) + \alpha_j' L_j'(\vec{x}) + \alpha_k' L_k'(\vec{x}) = 0.$$

Hence $G$ sends elements of $\mathcal{E}$ to elements of $\mathcal{E}$, we denote this action by

$$\Theta : \begin{cases} G \times \mathcal{E} & \longrightarrow & \mathcal{E} \\ (\sigma, E) & \longmapsto & \sigma(E) \end{cases}$$

So if we solve the $S$-unit equation corresponding to an element $E \in \mathcal{E}$ then we get the solutions to the $S$-unit equation corresponding to $\sigma(E)$ for free. This leads us to the following important observation:

LEMMA X.6. *It is enough when solving a TCDF equation to solve one $S$-unit equation from each orbit of the action, $\Theta$, of $G$ on $\mathcal{E}$.*

For instance consider a Thue equation of degree 4 such that the corresponding Galois group, $G$, contains a cyclic subgroup of order 4. We represent the element of order 4 in $G$ by $\sigma = (1, 2, 3, 4)$ where the indices denote a given ordering of the roots of $F(X, 1)$. Each linear form in the set $\mathcal{L}$ corresponds to one such root. The hyper-graph contains 4 hyper-edges which we can denote by, with an obvious notation,

$$E_1 = \{L_2, L_3, L_4\}, \quad E_2 = \{L_1, L_3, L_4\},$$
$$E_3 = \{L_1, L_2, L_4\}, \quad E_4 = \{L_1, L_2, L_3\}.$$

We need to consider the action of $\sigma$ on these hyper-edges:

$$\sigma(E_1) = \{\sigma(L_2), \sigma(L_3), \sigma(L_4)\} = \{L_3, L_4, L_1\} = E_2.$$

Similarly we see $\sigma(E_2) = E_3$ and $\sigma(E_3) = E_4$, hence the number of orbits is equal to one. So we need solve only one $S$-unit equation.

**X.3.3. Simplification of the unit equations.** Remember what we observed before in the proof of Theorem X.5. From

$$\prod_{j=1}^{m} L_j(\vec{x}) = C p_1^{z_1} \cdots p_t^{z_t}$$

and the unique factorization of ideals there is a finite number of possibilities for ideals $\mathfrak{a}_j$ such that

$$L_j(\vec{x})\mathcal{O}_S^* = \mathfrak{a}_j \mathcal{O}_S^*.$$

But if we then have an equation of the form

$$\alpha_1 L_1(\vec{x}) + \alpha_2 L_2(\vec{x}) + \alpha_3 L_3(\vec{x}) = 0$$

we can write $\alpha_j L_j(\vec{x}) = \alpha_j \beta_j \epsilon_j = \gamma_j \epsilon_j$, where $\beta_j$ come from a finite set and $\epsilon_j \in \mathcal{O}_S^* \cap \mathcal{O}_K$, to obtain

$$\gamma_1 \epsilon_1 + \gamma_2 \epsilon_2 + \gamma_3 \epsilon_3 = 0.$$

Hence from the earlier algorithm for $S$-unit equations we can obtain a finite set of possibilities for $(\epsilon_1/\epsilon_3, \epsilon_2/\epsilon_3)$, i.e. applying the algorithm with the groups $H_1 = H_2 = \mathcal{O}_S^*$. From this finite set of solutions we can determine a finite set of possibilities for $L_j(\vec{x})$ up to multiplication by an element of $\mathcal{O}_S^*$.

As we mentioned in Chapter IX the main thing to worry about when solving an $S$-unit equation is the rank of the two groups $H_1$ and $H_2$. However, in any reasonable example we may come across, the rank of $H_i = \mathcal{O}_S^*$ will be reasonably large ('large' with present computing machinery is around 5 or 6). We would therefore like some way of restricting our variables to range over subgroups of $H_1$ and $H_2$ of much smaller rank.

Here the Galois group can come in handy again. We have

$$\tau_i = \epsilon_i/\epsilon_3 = \frac{L_i(\vec{x})\beta_3}{L_3(\vec{x})\beta_i} \in H_i.$$

There are three main possibilities which can occur:

1. If $\tau_i$ is fixed by some subgroup of $G$ then we then know that $\tau_i$ is restricted to range over the $S$-units of a smaller number field than $K$. Hence the rank of the group $H_i$ can be made smaller.
2. Related to the above one may find that a subgroup of $G$ sends $\tau_i$ to $\tau_i^{-1}$. This gives linear equations on the exponents on the generators of $H_i$. Hence the rank of $H_i$ can again be reduced.
3. Sometimes one finds that there is a subgroup of $G$ which sends $\tau_1$ to $\tau_2^{\pm 1}$. This gives linear equations between the exponential variables in our unit equation.

Using these observations one can often significantly reduce the amount of effort needed in solving the relevant $S$-unit equations, see for instance [**181**].

## X.4. Exercises

1). Show that every binary form of degree greater than two with non-zero discriminant is triangularly connected, Lemma X.1.

2). Prove Lemma X.2 that $v \leq mn$.

CHAPTER XI

# Discriminant form equations

We shall now turn our attention to a special type of TCDF equation, namely discriminant and index form equations. Discriminant forms are an important example of TCDF equations. We shall first look at the general case of solving discriminant forms which arise from some number field of arbitrary degree ($\geq 3$). Then we shall consider special cases, in particular discriminant forms arising from quartic number fields. For such equations there is a rather nice algorithm due to Gaál, Pethő and Pohst which we shall explain.

## XI.1. Discriminant and index forms

A discriminant form is defined as follows: Let $1, \alpha_1, \ldots, \alpha_m$ denote $m+1$ linear independent algebraic integers and let $K = \mathbb{Q}(\alpha_1, \ldots, \alpha_m)$. We then define the linear form $\ell(\vec{x})$ by

$$\ell(\vec{x}) = \sum_{i=1}^{m} \alpha_i x_i.$$

The discriminant form of $\ell(\vec{x})$, with respect to $\mathbb{Q}$, is then the form

$$D_{K/\mathbb{Q}}(\ell(\vec{x})) = \prod_{1 \leq i < j \leq n} \left(\ell^{(i)}(\vec{x}) - \ell^{(j)}(\vec{x})\right)^2,$$

where $n = [K : \mathbb{Q}] \geq 3$. Clearly a discriminant form is a form of degree $n(n-1)$ in $m$ variables with coefficients in $\mathbb{Q}$. By definition it is certainly a decomposable form and we see that it is triangularly connected by noticing that for all $i, j, k$ with $i \neq j \neq k \neq i$ and $1 \leq i, j, k \leq n$ we have

$$\left(\ell^{(i)}(\vec{x}) - \ell^{(j)}(\vec{x})\right) + \left(\ell^{(j)}(\vec{x}) - \ell^{(k)}(\vec{x})\right) + \left(\ell^{(k)}(\vec{x}) - \ell^{(i)}(\vec{x})\right) = 0.$$

A discriminant form equation of Thue type is an equation where we wish to find all $\vec{x} \in \mathbb{Z}^m$ such that

$$D_{K/\mathbb{Q}}(\ell(\vec{x})) = D$$

for some given fixed integer $D$. A discriminant form equation of Mahler type is, not surprisingly, an equation where we wish to find solutions $\vec{x} \in \mathbb{Z}^m$ such that $\gcd(\vec{x}) = 1$ and

$$D_{K/\mathbb{Q}}(\ell(\vec{x})) = D p_1^{z_1} \cdots p_t^{z_t} \tag{XI.1}$$

for some given fixed set of prime numbers $p_1, \ldots, p_t$.

The elements $1, \alpha_1, \ldots, \alpha_m$ above clearly form a submodule of $\mathcal{O}_K$ of rank $m+1$. It can be rather inconvenient to consider such objects so we shall at once reduce the above problem to considering discriminant form equations of 'equation orders'. We put $K = \mathbb{Q}(\theta)$ for some algebraic integer $\theta$. The order $\mathbb{Z}[\theta]$ is called an 'equation order', for obvious reasons. It has as basis $1, \theta, \dots, \theta^{n-1}$ and we can consider the linear form

$$L(\vec{y}) = \sum_{i=2}^{n} y_i \theta^{n-1}$$

and the associated discriminant form

$$D_{K/\mathbb{Q}}(L(\vec{y})).$$

With the next lemma we reduce the study of the discriminant form $D_{K/\mathbb{Q}}(\ell(\vec{x}))$ to the discriminant form $D_{K/\mathbb{Q}}(L(\vec{y}))$.

LEMMA XI.1. *If we write*

$$\alpha_i = \sum_{j=1}^{n} a_{i,j} \theta^{j-1}$$

*for some $a_{i,j} \in \mathbb{Q}$ and set d to be the least common multiple of the denominators of the $a_{i,j}$ then we can deduce all the solutions to equation (XI.1) from the solutions of the equation*

$$D_{K/\mathbb{Q}}(L(\vec{y})) = d^{n(n-1)} D p_1^{z_1} \cdots p_t^{z_t}. \tag{XI.2}$$

PROOF. Let $A$ denote the matrix $(a_{i,j})$ and let $\vec{x}$ denote a solution to equation (XI.1). We define $\vec{y}$ by

$$\vec{y} = dA\vec{x}$$

and so $\vec{y} \in \mathbb{Z}^{n-1}$ and

$$L(\vec{y}) = d\ell(\vec{x}).$$

Hence

$$D_{K/\mathbb{Q}}(L(\vec{y})) = d^{n(n-1)} D_{K/\mathbb{Q}}(\ell(\vec{y})) = d^{n(n-1)} D p_1^{z_1} \cdots p_t^{z_t},$$

which is what we wanted to show. □

Often when talking about discriminant form equations authors sometimes refer to index form equations. In fact they are really one and the same thing (discriminant forms just having an added constant multiplier in them). Clearly from the linear form, $L(\vec{y})$, we see that $\theta^{(i)} - \theta^{(j)}$ divides $L^{(i)}(\vec{y}) - L^{(j)}(\vec{y})$ for all $i$ and $j$. But this then means that the discriminant of $\theta$,

$$D_{K/\mathbb{Q}}(\theta) = \prod_{1 \le i < j \le n} \left(\theta^{(i)} - \theta^{(j)}\right)^2,$$

must divide $D_{K/\mathbb{Q}}(L(\vec{y}))$. So we could assume that $D_{K/\mathbb{Q}}(\theta)$ divides $d^{n(n-1)}D$ in our equation above. When we take out this factor from the discriminant form we obtain the square of another form called the index form:

$$D_{K/\mathbb{Q}}(L(\vec{y})) = D_{K/\mathbb{Q}}(\theta) Ind^2_{K/\mathbb{Q}}(L(\vec{y})).$$

We end this introduction to discriminant forms by considering a very special case. If $m+1=n$ and $1, \alpha_2, \ldots, \alpha_n$ denotes an integral basis of the number field $K$ then the discriminant form $D_{K/\mathbb{Q}}(\ell(\vec{x}))$ always contains a constant factor which is equal to the discriminant of the field $K$, $D_K$. The other factor is a square of an index form. Solving the equation

$$D_{K/\mathbb{Q}}(\ell(\vec{x})) = D_K Ind^2(\ell(\vec{x})) = D_K$$

is equivalent to finding an element of index one in $K$, that is an element $\theta \in \mathcal{O}_K$ such that $\mathbb{Z}[\theta] = \mathcal{O}_K$.

This application of finding elements of given discriminant in a number field is the main application of discriminant forms. Equivalently discriminant form equations find monic irreducible polynomials of given discriminant with a root in a given field. Finding all monic irreducible polynomials of given degree and discriminant is then an algorithmic task as one can determine the possible discriminants of the associated fields, a fact that was first realized by Győry, see [**86**], [**87**], [**88**] and [**94**]. Fields of given degree and bounded discriminant are finite in number. Indeed there is an algorithm to determine all such fields [**153**]. See [**126**], [**128**] and [**129**] for examples of finding polynomials with given discriminant.

## XI.2. The general case: discriminant forms as TCDFs

We can, as discriminant forms are TCDFs, apply the method in the previous chapter to solve them. The first place in the literature where this was actually carried out in a given set of examples seems to have been in [**69**], where the case of discriminant forms of Thue type in biquadratic number fields was considered. In [**178**] a discriminant form of Mahler type in a quartic number field was also completely solved.

To apply the method we first we have to consider how many $S$-unit equations will we need to solve and how easy can we make such equations. In what follows we shall set

$$\gamma_{i,j,k} = \frac{L^{(i)}(\vec{y}) - L^{(j)}(\vec{y})}{L^{(k)}(\vec{y}) - L^{(i)}(\vec{y})}$$

and $K_{i,j,k} = \mathbb{Q}(\theta^{(i)}, \theta^{(j)}, \theta^{(k)})$. Then we can find a finite set of $\alpha_{i,j,k}$ such that we can write

$$\gamma_{i,j,k} = \alpha_{i,j,k}\tau_{i,j,k}$$

with $\tau_{i,j,k}$ an element of some finitely generated subgroup of $K^*_{i,j,k}$. Our $S$-unit equations are then

$$\alpha_{i,j,k}\tau_{i,j,k} + \alpha_{k,j,i}\tau_{k,j,i} + 1 = 0. \tag{XI.3}$$

Given a solution to one such equation we can deduce a solution to all other equations with the same indices (but permuted) using the following identities, which follow from the definition of $\gamma_{i,j,k}$:

$$\begin{array}{ll} \gamma_{i,k,j} = \gamma_{i,j,k}^{-1}, & \gamma_{k,i,j} = \gamma_{k,j,i}^{-1}, \\ \gamma_{j,i,k} = -1 - \gamma_{k,j,i}^{-1}, & \gamma_{j,k,i} = -1 - \gamma_{i,j,k}^{-1}. \end{array}$$

So we have at most ${}^nC_3/2$ such equations to solve. In Chapter X we suggested using the Galois group to reduce the number of such possibilities. This is what we shall now do. Let $K^{Gal}$ denote the minimal normal closure of $K$ and put $G = Gal(K/\mathbb{Q})$. From the definition of $\gamma_{i,j,k}$ an element $\sigma \in G$ acts on $\gamma_{i,j,k}$ by acting on the indices, i.e.

$$\sigma(\gamma_{i,j,k}) = \gamma_{\sigma(i),\sigma(j),\sigma(k)},$$

where we consider $\sigma$ as a permutation on some fixed ordering of the conjugates of $\theta$.

Each $S$-unit equation can be labelled by the ordered triple of indices of its first term. Hence equation (XI.3) above would be given the label $[i, j, k]$. But we have already noticed that given the solutions with respect to one such triple we can determine all the solutions with respect to any permutation of the triple. So we are really only interested in such triples up to permutation by elements of $S_3$. Hence we label each $S$-unit equation by a **set** $\{i, j, k\}$ which represents three such equations. The set of all such triples we shall denote by $R_n$, which will have ${}^nC_3$ members. Our task is then to deduce, as we mentioned in Chapter X, the number of orbits of $G$ on $R_n$ under the action

$$\begin{array}{ccc} G \times R_n & \longrightarrow & R_n, \\ (\sigma, \{i, j, k\}) & \longmapsto & \{\sigma(i), \sigma(j), \sigma(k)\}. \end{array}$$

For instance if it is the case that $K$ is a cyclic quintic extension, i.e. $n = 5$ and $G = C_5$, then the number of orbits is equal to 2.

We also suggested earlier that one could use the Galois group to make the rank of $H_1$ and $H_2$ much smaller. It is easy to see how to do this in the current situation. In what follows suppose that $\sigma'$ is a permutation of $\{1, \ldots, n\}$ which fixes $i, j$ and $k$, three arbitrarily chosen distinct indices. There are a variety of cases which come to our rescue and allow us to reduce the ranks, or to deduce linear equations between the exponents. Two examples which come to mind are

1. Where there exists a $\sigma$ of the form $\sigma = (i, k)\sigma' \in G$, where $\sigma'$ is a permutation which fixes $i, j$ and $k$. In such a situation we have $\sigma(\gamma_{i,j,k}) = -1 - \gamma_{i,j,k} = \gamma_{k,j,i}$. Hence we can express the exponents in $\gamma_{k,j,i}$ in terms of the exponents in $\gamma_{i,j,k}$. This gives us linear equations between the exponents as required.
2. Where there exists a $\sigma$ of the form $\sigma = (k, j)\sigma' \in G$, with $\sigma'$ as above. In this case we have $\sigma(\gamma_{i,j,k}) = \gamma_{i,j,k}^{-1}$. This will give a set of linear equations amongst the exponents in the $\gamma_{i,j,k}$-term. Usually one can then deduce that $\gamma_{i,j,k}$ must range over a group of smaller rank.

Clearly other cases can arise; see [**181**] for more details and examples.

## XI.3. A discriminant form equation in a cyclic quintic field

Let $K = \mathbb{Q}(\theta)$ denote the field considered in our example of Chapter IX, i.e.

$$\theta^5 + \theta^4 - 4\theta^3 - 3\theta^2 + 3\theta + 1 = 0.$$

We wish to compute all elements of index one in this field, in other words all elements $\alpha \in \mathcal{O}_K$ for which $\mathbb{Z}[\alpha] = \mathcal{O}_K$. Remember we have already remarked that $\mathcal{O}_K = \mathbb{Z}[\theta]$, so we can set

$$\alpha = l^{(i)}(\vec{x}) = x_1\theta + x_2\theta^2 + x_3\theta^3 + x_4\theta^4$$

where $x_i \in \mathbb{Z}$. Clearly we are only interested in $\alpha$ up to translations by an integer and scalings by $\pm 1$. If we then define

$$\gamma_{i,j,k} = \frac{l^{(i)} - l^{(j)}}{l^{(k)} - l^{(i)}} = \frac{\theta^{(i)} - \theta^{(j)}}{\theta^{(k)} - \theta^{(i)}}\epsilon_{i,j,k},$$

where $\epsilon_{i,j,k} \in \mathcal{O}_K^*$, we then have the $S$-unit equation

$$\gamma_{i,j,k} + \gamma_{k,j,i} + 1 = 0.$$

From our earlier discussion, owing to the action of the Galois group of $K$ we need only consider two such $S$-unit equations: one corresponding to $\{i, j, k\} = \{1, 2, 3\}$ and one corresponding to $\{i, j, k\} = \{1, 3, 4\}$.

If $Gal(K/\mathbb{Q}) = \langle\sigma\rangle$, where $\sigma = (1, 2, 3, 4, 5)$ as a permutation of the roots, we obtain the following ordering of the conjugates of $\theta$:

$$\begin{aligned} \theta^{(1)} &= \theta, \\ \theta^{(2)} &= -(\theta^4 + \theta^3 - 3\theta^2 - 2\theta + 1), \\ \theta^{(3)} &= \theta^3 - 3\theta, \\ \theta^{(4)} &= \theta^4 - 4\theta^2 + 2, \\ \theta^{(5)} &= \theta^2 - 2. \end{aligned}$$

As before we choose the four fundamental units to be

$$\eta_1 = \theta^2 + \theta - 1, \qquad \eta_2 = \theta^4 + \theta^3 - 3\theta^2 - 3\theta,$$
$$\eta_3 = \theta^3 + \theta^2 - 2\theta - 1, \qquad \eta_4 = \theta^4 - 3\theta^2 + 1.$$

We then find

$$\frac{\theta^{(1)} - \theta^{(2)}}{\theta^{(3)} - \theta^{(1)}} = -\eta_1\eta_4, \quad \frac{\theta^{(3)} - \theta^{(2)}}{\theta^{(1)} - \theta^{(3)}} = -\eta_1\eta_2\eta_3^{-1},$$

$$\frac{\theta^{(1)} - \theta^{(3)}}{\theta^{(4)} - \theta^{(1)}} = \eta_1^{-2}\eta_2^{-1}\eta_3\eta_4^{-1}, \quad \frac{\theta^{(4)} - \theta^{(3)}}{\theta^{(1)} - \theta^{(4)}} = \eta_1^{-2}\eta_2^{-1}\eta_3.$$

Hence our two $S$-unit equations can both be reduced to solving

$$\tau_1 + \tau_2 + 1 = 0$$

where $\tau_i \in \mathcal{O}_K^*$. But this was exactly the equation which was solved in Chapter IX. Remember that we found 570 solutions to this equation.

We look through all the possible exceptional units twice, once to determine $\gamma_{1,2,3}$ and once to determine $\gamma_{1,3,4}$. We then try to determine a non-trivial solution to the following four linear equations in $(x_1, x_2, x_3, x_4)$:

$$\begin{aligned} l^{(1)} - l^{(2)} - \gamma_{1,2,3}(l^{(3)} - l^{(1)}) &= 0, \\ l^{(3)} - l^{(2)} - (-1 - \gamma_{1,2,3})(l^{(1)} - l^{(3)}) &= 0, \\ l^{(1)} - l^{(3)} - \gamma_{1,3,4}(l^{(4)} - l^{(1)}) &= 0, \\ l^{(4)} - l^{(3)} - (-1 - \gamma_{1,3,4})(l^{(1)} - l^{(4)}) &= 0, \end{aligned}$$

These are easily solved by equating coefficients of powers of $\theta$ in the four equations. It turns out that the null space of the resulting systems always has dimension at most one. We find 55 such possible solutions $(x_1, x_2, x_3, x_4)$ which we then scale to have integer coprime values. We then test each of these 55 solutions to determine which give rise to elements $\alpha$ of index one. We find that there are 25 of these in all, which we now list as

$$\alpha = x_1\theta + x_2\theta^2 + x_3\theta^3 + x_4\theta^4$$

where $(x_1, x_2, x_3, x_4)$ is chosen from the following list:

$$\begin{aligned} &(-2,-3,1,1),\ (-3,1,1,0),\ (-3,0,1,0),\ (-2,-1,2,1),\ (0,1,0,0), \\ &(-2,-1,1,0),\ (-2,-4,1,1),\ (-11,-5,4,2),\ (-5,-13,2,3), \\ &(5,-11,-1,3),\ (-4,4,1,-1),\ (0,-4,0,1),\ (0,-3,0,1),\ (1,1,0,0), \\ &(-3,-3,1,1),\ (-2,-8,1,2),\ (1,-4,0,1),\ (1,0,0,0),\ (-1,2,1,-1), \\ &(-2,0,1,0),\ (-3,-4,1,1),\ (-5,-2,2,1),\ (-2,1,1,0),\ (1,-3,0,1), \\ &(1,-1,-1,0). \end{aligned}$$

So any element of index one must be of the form

$$t \pm \alpha$$

where $t$ is arbitrary and $\alpha$ comes from the above list.

## XI.4. Special cases

In this section we consider a variety of simple cases of small degrees. The easiest case is finding integral elements of given index/discriminant in cubic number fields. This problem is equivalent to solving a Thue equation over the integers, an observation which we leave as an exercise. There are many examples in the literature of solving discriminant/index form equations with cubic number fields both of Thue and Thue–Mahler type, see for instance [**74**] and [**128**].

**XI.4.1. Quartic number fields.** A simple method for this case has been developed by Gaál, Pethő and Pohst [**70**] and [**71**]. Their method is to reduce the problem to a cubic Thue equation and a pair of quadratic forms. These quadratic forms are themselves then reduced to Thue equations.

Let $K = \mathbb{Q}(\theta)$ denote a quartic number field defined by the equation

$$f(\theta) = \theta^4 + a_1\theta^3 + a_2\theta^2 + a_3\theta + a_4 = 0.$$

As we mentioned earlier the problem of computing elements of $\mathcal{O}_K$ of given discriminant can be reduced to the computation of elements of $\mathbb{Z}[\theta]$ with given discriminant. Say we wish to compute all $\alpha \in \mathbb{Z}[\theta]$ with discriminant given by $d$. If we write

$$\alpha = a + x\theta + y\theta^2 + z\theta^3,$$

we then have to determine all $(x, y, z) \in \mathbb{Z}^3$ such that $D_{K/\mathbb{Q}}(\alpha) = d$. We have

$$\prod_{i \neq j} \left( \frac{\alpha^{(i)} - \alpha^{(j)}}{\theta^{(i)} - \theta^{(j)}} \right) = \frac{d}{D_{K/\mathbb{Q}}(\theta)}. \tag{XI.4}$$

The right hand side of the above equation we shall denote by $i_d$; it is the square of the index of $\alpha$ in $\mathbb{Z}[\theta]$. Set

$$\xi_{i,j,k,l} = \theta^{(i)}\theta^{(j)} + \theta^{(k)}\theta^{(l)}.$$

This should ring lots of bells for anyone who has done an old fashioned course in Galois theory. This is nothing but the transformation one does when reducing the computation of the Galois group of a quartic polynomial to the Galois group of the 'resolvant cubic'. Indeed the resolvant cubic is given by $F(u, 1)$ where

$$\begin{aligned} F(u, v) &= (u - \xi_{1,2,3,4}v)(u - \xi_{1,3,2,4}v)(u - \xi_{1,4,2,3}v) \\ &= u^3 - a_2u^2v + (a_1a_3 - 4a_4)uv^2 + (4a_2a_4 - a_3^2 - a_1^2a_4)v^3. \end{aligned}$$

The crucial observation is the factors in the left hand side of equation (XI.4) can be grouped together (up to a factor of $\pm 1$) into three groups of two such that

$$\left( \frac{\alpha^{(i)} - \alpha^{(j)}}{\theta^{(i)} - \theta^{(j)}} \right) \left( \frac{\alpha^{(k)} - \alpha^{(l)}}{\theta^{(k)} - \theta^{(l)}} \right) = Q_1(x, y, z) - \theta_{i,j,k,l}Q_2(x, y, z)$$

for $(i, j, k, l) = (1, 2, 3, 4)$, $(1, 3, 2, 4)$ or $(1, 4, 2, 3)$ where the ternary quadratic forms $Q_1(x, y, z)$ and $Q_2(x, y, z)$ are given by

$$\begin{aligned} Q_1(x, y, z) &= x^2 - xya_1 + xz(a_1^2 - 2a_2) + yz(a_3 - a_1a_2) \\ &\quad + z^2(-a_1a_3 + a_2^2 + a_4), \\ Q_2(x, y, z) &= y^2 - xz - a_1yz + z^2a_2. \end{aligned}$$

Putting all the above together we have:

LEMMA XI.2. *Finding all elements $\alpha = a + x\theta + y\theta^2 + z\theta^3 \in \mathbb{Z}[\theta]$ with discriminant $d$ is equivalent to solving the system of equations*

$$\begin{aligned} F(u,v) &= \pm i_d, \\ Q_1(x,y,z) &= u, \\ Q_2(x,y,z) &= v. \end{aligned}$$

We have already seen how to solve Thue equations in Chapter VII. Hence we can easily determine a finite set of possible pairs $(u, v)$. All that remains is to solve the final two equations for $x, y$ and $z$. Before we proceed to this we pause to note that $F(u, v) = \pm i_d$ is trivial to solve if it is not irreducible which is equivalent to saying that $f(X)$ does not have Galois group $A_4$ or $S_4$. This is the Galois theory we mentioned above. In the non-trivial case it is only a cubic Thue equation so it really should not be that hard to solve.

We shall now describe how Gaál, Pethő and Pohst solve the final problem of finding $x, y$ and $z$ such that

$$Q_1(x,y,z) = u \ , \ Q_2(x,y,z) = v$$

for given integers $u$ and $v$. Firstly define a new quadratic form

$$Q_0(x,y,z) = vQ_1(x,y,z) - uQ_2(x,y,z).$$

Then any solution $(x, y, z)$ to our original pair will satisfy $Q_0(x, y, z) = 0$. Hence, as we have seen before in Chapters IV and VII, we can express the coprime integer solutions to $Q_0(x, y, z) = 0$ (if it has any) in terms of three binary quadratic forms in two unknown integer coprime variables:

$$gx = q_1(p,q) \ , \ gy = q_2(p,q) \ , \ gz = q_3(p,q)$$

where $q_i(p, q) \in \mathbb{Z}[p, q]$, where $g$ is an integer from some finite set. We now wish to determine $p$ and $q$, but this can be achieved by substituting the $q_i$ into the forms $Q_1$ and $Q_2$ to obtain two quartic Thue equations. It turns out (for a proof see [**71**]) that these quartic Thue equations are norm forms in two variables from the field $K$ down to $\mathbb{Q}$.

So to solve a quartic discriminant we have to solve one cubic Thue equation, $F(u, v) = \pm i_d$. Then we use these solutions to construct curves of genus zero, $Q_0(x, y, z) = 0$, which we can solve by the methods of Chapter IV. From a single solution to this curve of genus zero we then construct two quartic Thue equations which determine our solutions.

**XI.4.2. Sextic number fields.** We now jump to sextic number fields as not much work has been done on the special case of quintic number fields. When the sextic number field has a quadratic subfield then the discriminant form can be reduced to considering 'relative Thue equations'. These are Thue equations, in this case of degree 3, whose coefficients and variables come from the ring of integers of an algebraic number field, in this case the quadratic subfield. These Thue equations can be solved with a method similar to the one described in Chapter VII, or one could indeed use the method of Chapter

X as Thue equations are TCDF equations. For more details on the special case of sextic number fields see [**67**], [**66**] and [**72**].

**XI.4.3. Parametric Families.** In [**68**], Gaál determines all the elements of index one in five families of totally complex quartic number fields, including the fields $\mathbb{Q}(\theta)$ where $\theta^4 + k = 0$. For another parameterized set of quartic index form equations see [**132**] and [**119**]. In addition a parametric family of real cyclic quintic fields was considered in [**73**].

Gaál [**65**] has shown that if $K$ is a compositum of two number fields $M$ and $L$ with coprime discriminants then one can find elements of index one in $K$ by solving two smaller equations. This has allowed him to show that there are no elements of index one in a parametric family of number fields of degree ten. The basic result which allows this is the following:

THEOREM XI.3. *Let $M$ have integral basis $\{m_1 = 1, m_2, \ldots, m_s\}$ and discriminant $D_M$ and let $L$ have integral basis $\{l_1 = 1, l_2, \ldots, m_r\}$ and discriminant $D_L$. If $(D_L, D_M) = 1$ and*

$$\alpha = \sum_{i=1}^{r}\sum_{j=1}^{s} x_{i,j} l_i m_j$$

*is an element of index one then $x_{i,j} \in \mathbb{Z}$ and*

$$\begin{aligned} N_{M/\mathbb{Q}}\left(I_L\left(\sum_{i=1}^{s} x_{2,i}m_i, \ldots, \sum_{i=1}^{s} x_{r,i}m_i\right)\right) &= \pm 1, \\ N_{L/\mathbb{Q}}\left(I_M\left(\sum_{i=1}^{r} x_{i,2}l_i, \ldots, \sum_{i=1}^{r} x_{i,s}l_i\right)\right) &= \pm 1, \end{aligned}$$

*where $I_L(a_2, \ldots, a_r)$ and $I_M(b_2, \ldots, b_s)$ are the index forms, over $L$ and $M$ respectively, of the linear forms*

$$\sum_{i=2}^{r} a_i l_i \ , \ \sum_{i=2}^{s} b_i m_i.$$

**XI.4.4. Computing all power integral bases of a specific quartic number field.** We end this chapter by using the method outlined above for solving index form/discriminant form equations in quartic number fields to determine all power integral bases of the field $K = \mathbb{Q}(\theta)$, where $\theta^4 + 1 = 0$ This means that we wish to determine all elements of $\mathcal{O}_K = \mathbb{Z}[\theta]$ of index one. We write $\alpha$ in the form

$$\alpha = a + x\theta + y\theta^2 + z\theta^3$$

where $a \in \mathbb{Z}$ is arbitrary and $x, y, z \in \mathbb{Z}$ are to be found. The values of $x, y$ and $z$ must satisfy the system of equations

$$\begin{aligned} F(u,v) = u^3 - 4uv^2 &= \pm 1, \\ Q_1(x,y,z) = x^2 + z^2 &= u, \\ Q_2(x,y,z) = y^2 - xz &= v. \end{aligned}$$

From the equation $F(u,v) = \pm 1$ we can see that we must have $(u,v) = (\pm 1, 0)$. Then from the equation $Q_2(x,y,z) = 0$ we see that, on applying the methods of Chapter IV,

$$gx = -p^2 \ , \ gy = p^2 - pq \ , \ gz = -p^2 + 2pq - q^2.$$

If we insist that $(x,y,z)$ are coprime integers and $(p,q)$ are also coprime integers then the only possibilities for $g$ are $\pm 1$. If we then substitute these formulae for $x, y$ and $z$ into $Q_1(x,y,z) = \pm 1$ we obtain the quartic Thue equation

$$2p^4 - 4qp^3 + 6q^2p^2 - 4q^3p + q^4 = \pm 1.$$

We find, using the methods of Chapter VII, that this Thue equation only has the solutions $(p,q) = \pm(1,1)$ and $\pm(0,1)$. In the first case we obtain $(x,y,z) = \pm(1,0,0)$, whilst in the second we obtain $(x,y,z) = \pm(0,0,1)$, which means that the only elements of index one in $K(\theta)$ are of the form

$$\alpha = a \pm \theta \text{ and } \alpha = a \pm \theta^3,$$

for any $a \in \mathbb{Z}$.

## XI.5. Exercises

1). Show that every discriminant form can be factored into the product of a fixed integer and the square of another form.

2). Show that the problem of computing integral elements of given index and/or discriminant in a cubic number field is equivalent to solving a Thue equation over the integers.

# Part 3

# Integral and rational points on curves

CHAPTER XII

# Rational points on elliptic curves

In previous chapters we have seen how to solve the problem of finding all integral points on an elliptic curve. This is only one of the two fundamental diophantine questions which one can ask about an elliptic curve. In this chapter we look at the other question: What is the structure of the set of rational points on an elliptic curve? We shall give a sketch of the proof that the set of rational solutions to an elliptic curve forms a finitely generated abelian group, known as the Mordell–Weil group. There is no known effective proof of this result; we shall, however, outline a possible algorithmic proof which works fine in practice most of the time.

We first consider the basic theory of elliptic curves, which we shall just skim over. Those of you who have not met any of this before should perhaps consult any one of the excellent textbooks in the area such as [**107**], [**172**], [**101**], [**25**], [**105**], [**176**] or [**40**]. You could also, perhaps, consult the survey articles [**22**] and [**194**]. We shall only consider those parts of the theory of elliptic curves which we need in this book.

After outlining the basic theory we shall outline two 'algorithms' for determining $E(\mathbb{Q})/2E(\mathbb{Q})$. These two methods should not really be called algorithms as they are not guaranteed to work. However, in practice they often do and when they do not we can apply further ideas which in most cases complete the desired task. We shall then outline a method for performing the 'infinite descent'. At this stage we use the generators we have just determined for $E(\mathbb{Q})/2E(\mathbb{Q})$ to determine generators for $E(\mathbb{Q})$.

## XII.1. Basics on elliptic curves

An elliptic curve is a non-singular curve of the form

$$E : Y^2 + a_1XY + a_3Y = X^3 + a_2X^2 + a_4X + a_6,$$

where $a_i \in K$, some number field. Non-singular means that for no point on the curve do the partial differentials, in the $X$ and $Y$ directions, simultaneously vanish. In such a situation the curve is said to be defined over $K$. An elliptic curve of the above form (long Weierstrass form) has at least one rational point, namely the point at infinity. You should think of the point at infinity as a point infinitely far up the $y$-axis or equivalently the point which has projective coordinates $(0, 1, 0)$.

We are interested in determining the structure of all the rational points on $E$. From the equation for $E$ we can define the auxiliary quantities which will come in handy later:

$$\begin{aligned} b_2 &= a_1^2 + 4a_2, \\ b_4 &= a_1 a_3 + 2a_4, \\ b_6 &= a_3^2 + 4a_6, \\ b_8 &= (b_2 b_6 - b_4^2)/4, \\ c_4 &= b_2^2 - 24b_4, \\ c_6 &= -b_2^3 + 36b_2 b_4 - 216b_6. \end{aligned}$$

In addition to these we define the discriminant to be $\Delta = (c_4^3 - c_6^2)/1728$ and the $j$-invariant to be $j = c_4^3/\Delta$. These last two constants determine fundamental properties of the elliptic curve. The curve is non-singular if and only if $\Delta \neq 0$, while the $j$ invariant classifies elliptic curves up to isomorphism over the complex numbers.

An isomorphism of an elliptic curve is a mapping which keeps the point at infinity fixed. Such a map, from a curve in long Weierstrass form to another in long Weierstrass form, must be of the form

$$x = u^2x' + r\ ,\ y = u^3y' + su^2x' + t,$$

where $r, s, t \in K$ and $u \in K^*$. Such an isomorphism keeps $j$ fixed and multiplies $c_4$ by $u^{-4}$, $c_6$ by $u^{-6}$ and $\Delta$ by $u^{-12}$. If $K = \mathbb{Q}$ then we can find a curve with integer coefficients in the isomorphism class of any given curve which has the smallest possible value for $\Delta$. We call such a curve a 'minimal model' for our original curve and the resulting smallest possible value for $\Delta$ we call the minimal discriminant.

With an isomorphism of the above form we can make $E$ isomorphic to the curve

$$Y^2 = 4X^3 + b_2X + 2b_4X + b_6$$

or even to the curve

$$Y^2 = X^3 - 27c_4X - 54c_6 = X^3 + AX + B.$$

Putting $g_2 = c_4/12$ and $g_3 = c_6/216$ we see that $E$ is isomorphic to

$$Y^2 = 4X^3 - g_2X - g_3,$$

which should ring bells for anyone who has seen elliptic functions before. Indeed it is the link between a general elliptic curve and elliptic functions which provides the method of the next chapter for solving the integral point problem.

An elliptic curve can have at most two real components depending on whether the discriminant $\Delta$ is positive or negative. We shall call the unbounded component the identity component, or $E^0(\mathbb{R})$; it is the part of the curve in the real plane which is 'unbounded' or 'connected to $\infty$'. The other possible component (which only occurs in the case $\Delta > 0$) we shall call the 'egg'. This is because it looks like an 'egg' when you actually draw the curve.

For the rest of this section we shall concentrate on the structure of the $K$-rational points on such a curve. We denote by $E(K)$ the set of $K$-rational points on the curve $E$ which is assumed to be defined over $K$, i.e. $E(K)$ is the set of all points with coordinates in the algebraic number field, $K$. We can obviously assume our curve is in short Weierstrass form,

$$E : Y^2 = X^3 + AX + B, \qquad \text{(XII.1)}$$

as a change of variable from long Weierstrass form to short Weierstrass form is a bijection from the set of $K$-rational points to the set of $K$-rational points. Let $P, Q$ be two distinct $K$-rational points on $E$. We can join $P$ and $Q$ by a straight line which must intersect $E$ at one further point, say $R$. We then reflect $R$ in the $x$-axis and call the resulting point $P + Q$ (see Figure XII.1 for an example). A moment's thought will reveal that $R$ and hence $P + Q$ must also be $K$-rational points.

To add $P$ to itself we take the tangent to the curve at $P$. Such a line must intersect $E$ in one other point, say $R$, as $E$ is defined by a cubic equation. Again we reflect $R$ in the $x$-axis to obtain a point which we call $[2]P = P+P$.

The above process of determining $P + Q$ given $P$ and $Q$ is often called the 'chord–tangent process'. The operation on points which we have just explained can be shown to define an additive abelian group law on $E(K)$ with the point at infinity as the zero. To sum up:

**Three points sum to zero in the group if there is a straight line joining them.**

The resulting group is called the Mordell–Weil group of $E$. It is rather easy to determine explicit formulae for the above group law. We now give such formulae in terms of an elliptic curve in long Weierstrass form:

LEMMA XII.1. *Let $E$ denote an elliptic curve given by*

$$E : Y^2 + a_1XY + a_3Y = X^3 + a_2X^2 + a_4X + a_6$$

*and let $P_1 = (x_1, y_1)$ and $P_2(x_2, y_2)$ denote points on the curve. Then*

$$-P_1 = (x_1, -y_1 - a_1x_1 - a_3).$$

*Set*

$$\lambda = \frac{y_2 - y_1}{x_2 - x_1}, \quad \mu = \frac{y_1x_2 - y_2x_1}{x_2 - x_1}$$

*when $x_1 \neq x_2$ and*

$$\lambda = \frac{3x_1^2 + 2a_2x_1 + a_4 - a_1y_1}{2y_1 + a_1x_1 + a_3}, \quad \mu = \frac{-x_1^3 + a_4x_1 + 2a_6 - a_3y_1}{2y_1 + a_1x_1 + a_3}$$

*when $x_1 = x_2$. If*

$$P_3 = (x_3, y_3) = P_1 + P_2$$

FIGURE XII.1. The group law on an elliptic curve

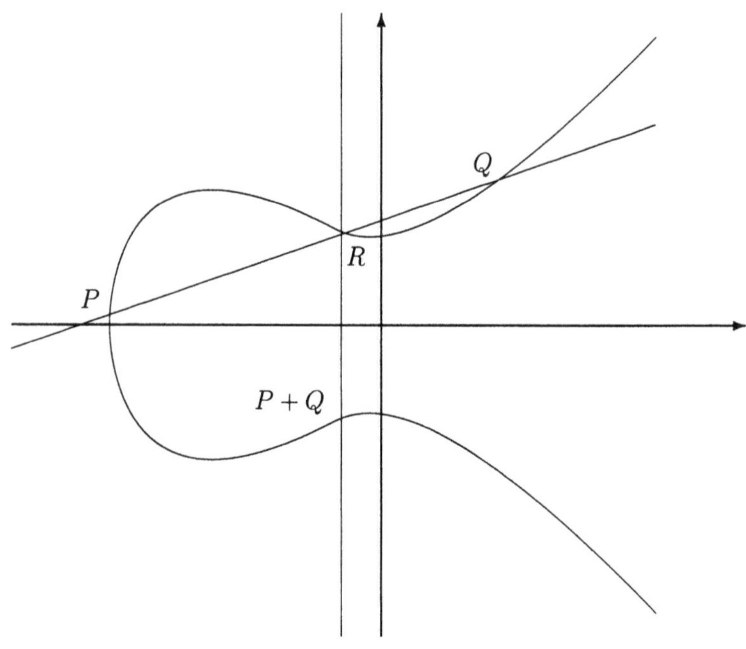


*then $x_3$ and $y_3$ are given by the formulae*

$$\begin{aligned} x_3 &= \lambda^2 + a_1\lambda - a_2 - x_1 - x_2, \\ y_3 &= -(\lambda + a_1)x_3 - \mu - a_3. \end{aligned}$$

PROOF. Exercise □

The isomorphisms described earlier then become group isomorphisms as they respect the group structure. In the late 19th century, Poincaré conjectured that the group $E(\mathbb{Q})$ was finitely generated. This result was proved in 1922 by Mordell and later generalized to arbitrary number fields by Weil:

THEOREM XII.2 (Mordell–Weil). *The group $E(K)$ is finitely generated.*

We will also require a rudimentary knowledge of the canonical height on elliptic curves. This is derived from the heights on projective space which we met earlier in Section II.1. The (naive) height of a point on an elliptic curve is defined to be the height of the $x$-coordinate considered as a point in $\mathbb{P}^1(K)$.

I.e. if $P = (x, y) \in E(\mathbb{Q})$ and $x = n/d$ is a fraction in lowest terms then

$$h(P) = h(x(P)) = \max\{\log|n|, \log|d|\}.$$

This height function almost has the properties we require but it is slightly cumbersome to use in some situations. We therefore modify it using the following formula to produce a new height function called the Neron–Tate height (or the canonical height).

$$\hat{h}(P) = \lim_{N\to\infty} 4^{-N} h(2^N P). \tag{XII.2}$$

THEOREM XII.3 (Neron–Tate). *The Neron–Tate height has the following properties:*

- *For all $P, Q \in E(K)$,*
$$\hat{h}(P+Q) + \hat{h}(P-Q) = 2\hat{h}(P) + 2\hat{h}(Q).$$
- *For $P \in E(K)$ and $m \in \mathbb{Z}$ we have*
$$\hat{h}(mP) = m^2 \hat{h}(P).$$
- *$\hat{h}$ is a quadratic form on $E(K)$ and defines a pairing*
$$\langle P, Q\rangle = \frac{1}{2}\left(\hat{h}(P+Q) - \hat{h}(P) - \hat{h}(Q)\right)$$
*which is bilinear.*
- *$\hat{h}(P) \geq 0$ with equality if and only if $P$ is a torsion point.*

While (XII.2) gives a definition for $\hat{h}(P)$ this is not the most efficient way of computing the Neron–Tate height. Usually to compute the canonical height one decomposes it into a sum of local components (just as the naive height can be decomposed into a sum of local components). There are then fast methods of computing these local components, see for instance [**32**], [**40**], [**173**], [**175**] and [**221**]. One should be wary when using canonical heights as different authors use different normalizations (sometimes the right hand side of (XII.2) is divided by two). For instance the canonical height here is twice that of the canonical height in the papers by Silverman and the one used in the package **PARI**.

It is also possible to bound the difference between the canonical and the naive heights [**222**] and [**174**]:

LEMMA XII.4. *There are constants $c_1, c_2$ such that*

$$-c_1 \leq \hat{h}(P) - h(P) \leq c_2.$$

There is a procedure due to Siksek [**169**] which determines a very accurate value for the constant $c_1$, which is all that we shall need in this chapter. In the next chapter we shall also need a value for the constant $c_2$. In the case $K = \mathbb{Q}$ and when the elliptic curve is defined by a Weierstrass equation with integer coefficients we can take the following values for the above constants, see [**174**]:

$$c_1 = \frac{1}{12}h(j) + \mu(E) + 1.946 \text{ and } c_2 = \mu(E) + 2.14,$$

where

$$\mu(E) = \frac{\log|\Delta| + \log^+(j)}{6} + \log^+(b_2/12) + \log(2^*),$$

$\log^+(x) = \max\{1, \log|x|\}$ and $2^* = 2$ if $b_2 \neq 0$ and $2^* = 1$ otherwise.

Again these constants depend on the normalization for the canonical height we have chosen. Using the above we can easily compute all torsion points on $E(K)$.

LEMMA XII.5. *If $P = (X, Y) \in E(K)$ is of finite order then $h(X) \leq c_1$.*

PROOF. This follows from the above lemma as a torsion point will have zero canonical height. One can then bound the naive height using the above constants. □

Finally given a set of $r$ points, $P_1, \ldots, P_r$ on $E$ we can define the height pairing matrix which is given by $(\langle P_i, P_j \rangle)$. The determinant of this matrix will be zero if and only if the $r$ points are dependent. In addition if $P_1, \ldots, P_r$ are indeed generators of the free part of $E(K)$ then we can use the height pairing matrix to consider $E(K)$ as a lattice in $\mathbb{R}^r$ with a norm given by the canonical height. We have that if $P = n_1P_1 + \cdots + n_rP_r$ then

$$\hat{h}(P) = \vec{n}^t R \vec{n}.$$

The absolute value of the determinant of the height pairing matrix when the given points are generators of the free part of $E(K)$ is called the regulator of the elliptic curve, $R(E(K))$. It clearly does not depend on which choice of generators we happen to have made. The regulator of an elliptic curve plays an analogous role to that played by the regulator of a number field. Indeed the rational points on an elliptic curve behave in a way which resembles the unit group of a number field.

## XII.2. The weak Mordell–Weil theorem

In this section we shall concentrate on the following theorem:

THEOREM XII.6 (weak Mordell–Weil). *The group $E(K)/2E(K)$ is a finite group.*

What we really want is an algorithmic proof of this result, that is a proof which gives an algorithm to actually compute generators of $E(K)/2E(K)$. Alas there is no known algorithmic proof owing to the failure of the Hasse principle for curves of genus one. However, for many particular curves we can provide such an algorithmic proof. Although we shall concentrate on the case

$K = \mathbb{Q}$, similar techniques work for other number fields (however, they are more complicated). We shall assume our curve is in short Weierstrass form,

$$Y^2 = X^3 + AX + B,$$

as we are interested in rational solutions. If $E(\mathbb{Q})$ has a rational point of order 2, i.e. $X^3 + AX + B$ has a linear factor over $\mathbb{Q}$, then one could use the method of descent via 2-isogeny to determine $E(\mathbb{Q})/2E(\mathbb{Q})$, see [**172**]. There are two popular ways of trying to compute $E(\mathbb{Q})/2E(\mathbb{Q})$ which do not depend on the 2-torsion structure of $E(\mathbb{Q})$, both are methods are generally referred to as 2-descent.

**XII.2.1. Weak Mordell–Weil theorem: the direct method.** This is the most ancient method. We shall only skim over the method; for more details consult the book by Cassels [**25**]. We write $Y = y/z^3$ and $X = x/z^2$ where $(x, y, z) \in \mathbb{Z}^3$ with $(x, z) = 1$. Our equation then becomes

$$y^2 = x^3 + Axz^4 + Bz^6.$$

The idea is now to mimic what was done in Section VII.4 for integral points on elliptic curves. We let $L$ denote the algebra

$$L = \mathbb{Q}[T]/(T^3 + AT + B).$$

This algebra is a sum of at most three number fields, $L = \sum_{i=1}^{t} L_i$, and there is a map

$$\Theta : \begin{cases} E(\mathbb{Q}) & \longrightarrow & L^*/L^{*2} \\ P & \longmapsto & x(P) - Tz(P)^2 \end{cases}.$$

The important result is:

LEMMA XII.7. *The map $\Theta$ above is a group homomorphism with kernel $2E(\mathbb{Q})$.*

Hence if we can show that the image of $\Theta$ is finite then we will have proved the weak Mordell–Weil theorem. Let $S$ denote the set of primes which divide the discriminant of $E$ (we shall also throw the prime 2 into $S$ for good measure). Just as in the Chapter VII we find that the image of $\Theta$ is contained in

$$\sum_{i=1}^{t} L_i(S, 2).$$

But this is a sum of finite groups. Hence the image of $\Theta$ is finite and we have the required result.

However, we want an algorithmic proof, which means we need to explicitly determine generators of the image of $\Theta$. For ease of exposition we shall assume that $t = 1$, i.e. that $X^3 + AX + B$ is irreducible. The other cases are left as an exercise. In our case $L$ is itself a number field, and we have

$$X - \theta Z^2 = \alpha\beta^2 \qquad \text{(XII.3)}$$

where $\theta^3 + A\theta + B = 0$, $\alpha \in L(S, 2)$ and $\beta \in L$. We also know that

$$\mathrm{N}_{L/\mathbb{Q}}(\alpha) = \text{ square}.$$

Hence we are already restricted to a finite number of possible $\alpha$. For each of these $\alpha$ we need to determine whether a $\beta$ exists such that the above equation (XII.3) is soluble in integers $X$ and $Z$. In doing so we can make use of the fact that such all such $\alpha$ which have this additional property form a group (i.e. the image of $\Theta$).

We write $\beta = x_1 + x_2\theta + x_3\theta^2$, wishing to determine if there are rational numbers $x_1, x_2, x_3$, and rational integers $X, Z$ such that

$$X - \theta Z^2 = \alpha(x_1 + x_2\theta + x_3\theta^2)^2.$$

Expanding the right hand side we find three quadratic forms in $x_1, x_2, x_3$ such that

$$X - \theta Z^2 = Q_1(x_1, x_2, x_3) + Q_2(x_1, x_2, x_3)\theta + Q_3(x_1, x_2, x_3)\theta^2.$$

Equating coefficients of $\theta^i$ we obtain the three equations

$$\begin{aligned} Q_1(x_1, x_2, x_3) &= X, \\ Q_2(x_1, x_2, x_3) &= -Z^2, \\ Q_3(x_1, x_2, x_3) &= 0 \end{aligned}$$

We first look at the last equation. This is a curve of genus 0 and as such satisfies the Hasse principle. So we can easily determine whether it has any rational solutions at all, see Chapter IV. This leads to our first criterion:

**Criterion 1:** $Q_3(x_1, x_2, x_3) = 0$ **must be locally soluble everywhere.**

Given that it has rational solutions we can find one and then express all possible rational solutions $(x_1, x_2, x_3)$ in terms of three binary quadratic forms, as in Chapter IV,

$$x_1 = q_1(m, n)\ ,\ x_2 = q_2(m, n)\ ,\ x_3 = q_3(m, n),$$

where $m$ and $n$ are rational variables and the $q_i$ have rational coefficients (hence the non-appearance of the constant $g$ from Chapter IV). We then substitute these three quadratic forms into the second equation above to obtain

$$Z^2 = G(m, n), \tag{XII.4}$$

where $G(m, n) = -Q_2(q_1(m, n), q_2(m, n), q_3(m, n))$ is a binary quartic form. Such an equation will be called a 'quartic' for short in the following discussion. Clearing denominators we can assume that we wish to solve (XII.4) for integral values of $m, n$ and $Z$. Any solution to (XII.4) will lead to a solution to our original problem, i.e. an element of the image of $\Theta$. Likewise an element in the image of $\Theta$ will give rise to a solution to (XII.4).

So the only problem is to determine whether (XII.4) has any rational solutions and if it does to determine such a solution. Alas equation (XII.4) determines a curve of genus one, which we know may not satisfy the Hasse principle. Hence it may be the case that while the curve has points locally for every prime it has no global solution. In any case we can eliminate many obviously non-solutions with our second criterion:

**Criterion 2: Equation (XII.4) must be locally soluble everywhere.**

It can be shown (using some Galois cohomology) that all the $\alpha$ which pass both the above criterion form a finite group. This group is called the 2-Selmer group, $S_2$. It is clear that the image of $\Theta$ lies in $S_2$ and hence $S_2$ contains a subgroup isomorphic to $E(\mathbb{Q})/2E(\mathbb{Q})$.

By abuse of notation we shall think of elements of $S_2$ as being locally soluble equations of the form (XII.4) and elements of $E(\mathbb{Q})/2E(\mathbb{Q})$ as being those equations in $S_2$ which have a global solution. We have the exact sequence

$$0 \to E(\mathbb{Q})/2E(\mathbb{Q}) \to S_2 \to Ш_2 \to 0.$$

The group $Ш_2$ is the 2 torsion part of the Tate–Shaferevich group. It is the obstruction to us having an algorithm to determine $E(\mathbb{Q})/2E(\mathbb{Q})$.

In many examples we have $Ш_2 = \{0\}$. If $Ш_2 \neq \{0\}$ then we will have to come up with a way of determining which of our elements in $S_2$ have a global solution. Using the sieving technique of Section III.4 we can determine those elements of $S_2$ which have a global point of small height. There are many ways of proceeding for elements of $S_2$ which have no points of small height, for instance using higher descents, see [**19**], [**18**] and [**127**]. There are other techniques of showing the non-existence of rational points which can often help here too, see [**170**].

We shall now show that $E(\mathbb{Q})/2E(\mathbb{Q})$ has order 2 for the elliptic curve given by $Y^2 = X^3 - 2$. Following the method above we let $K = \mathbb{Q}(\theta)$ where $\theta^3 - 2 = 0$. Using a computer package such as **PARI** we can determine that the class number of $K$ is one and for the single fundamental unit of $K$ we can take $\theta - 1$. The relevant prime ideals we need to consider are those dividing 2 and 3, both of which completely ramify. As generators of the two prime ideals we can take $\theta$ and $1 + \theta$. We then find that $K(S, 2)$ has rank four and generators given by

$$-1\ ,\ \theta - 1\ ,\ \theta \text{ and } 1 + \theta.$$

However, we require the subgroup of $K(S, 2)$ which is contained in the kernel of the norm map from $K$ down to $\mathbb{Q}^*/\mathbb{Q}^{*2}$. It does not take too long to decide that this kernel can contain only one non-trivial element, namely $\theta - 1$.

We therefore wish to look at the equation

$$X - \theta Z^2 = (\theta - 1)(x_1 + x_2\theta + x_3\theta^2)^2.$$

Expanding and equating coefficients of powers of $\theta$ we obtain the simultaneous equations

$$\begin{aligned} Q_2(x_1, x_2, x_3) &= x_1^2 - 2x_1x_2 + 4x_3x_2 - 2x_3^2 &= -Z^2, \\ Q_3(x_1, x_2, x_3) &= 2x_2x_1 - 2x_3x_1 + 2x_3^2 - x_2^2 &= 0. \end{aligned}$$

The second of these equations we deal with using the techniques of Chapter IV. We first spot a single solution of $Q_3 = 0$, which we take as $(x_1, x_2, x_3) = (1, 0, 0)$. It is then a simple matter to deduce that every solution of $Q_3 = 0$ must be given by

$$x_1 = p^2 - 2q^2 \ , \ x_2 = 2p^2 - 2pq \ , \ x_3 = 2pq - 2q^2$$

where $p$ and $q$ range through all coprime integer values. Substituting these values into the equation $Q_2 = -Z^2$ we obtain the quartic

$$-Z^2 = -3p^4 + 20p^3q - 36p^2q^2 + 24pq^3 - 4q^4$$

which has the 'obvious' solution $(p, q, Z) = (0, 1, 2)$. From this we then see that the pair of quadratic forms has the simultaneous solution $(x_1, x_2, x_3, Z) = (-2, 0, -2, 2)$, which means that we can write

$$X - \theta Z^2 = (\theta - 1)(-2 - 2\theta^2)^2 = 12 - 4\theta.$$

So we find that $S_2 = E(\mathbb{Q})/2E(\mathbb{Q})$ has two elements, as a representative of the non-trivial element we can take the point $(3, 5) \in E(\mathbb{Q})$. In this case the proof of the weak Mordell–Weil theorem is completely effective. As the next example shows this could have been because we were lucky.

We shall now attempt to compute the Mordell–Weil group of the curve $Y^2 = X^3 - 141$, and try to deduce the structure of its group of rational points. We sketch the method leaving the details as an exercise. Letting $K = \mathbb{Q}(\theta)$ where $\theta^3 - 141 = 0$ we find that $CL_K \cong C_2 \times C_4$ and a fundamental unit of $K$ is given by $\eta = -253801 - 152888\theta + 123060\theta^2$. The prime ideals which we need to consider to compute $K(S, 2)$ are those lying above 2, 3 and 47:

$$(2) = \mathfrak{p}_1\mathfrak{p}_2 \ , \ (3) = \mathfrak{p}_3^3 \ , \ (47) = \mathfrak{p}_4^3.$$

If $\mathfrak{a}$ and $\mathfrak{b}$ are the generators of the two cyclic components of the class group then we find that we can take (with an explicit choice of $\mathfrak{a}$ and $\mathfrak{b}$),

$$\mathfrak{a}^2 = \langle \psi \rangle \ , \ \mathfrak{b}^4 = \langle \phi \rangle.$$

where $\psi = -1192 - 229\theta - 44\theta^2$ and $\phi = -5 + \theta$. A short calculation with a package like **PARI** allows us to deduce that $K(S, 2)$ in this example has rank 7 and as generators we can take

$$-1, \eta, -10 + 2\theta, 75 - 4\theta - 2\theta^2, -7802 - 1499\theta - 288\theta^2, \psi \text{ and } \phi.$$

From this group of order 128 we need only consider the subgroup which lies in the kernel of the norm map from $K^*/K^{*2}$ down to $\mathbb{Q}^*/\mathbb{Q}^{*2}$. This subgroup has rank 3 and a quick calculation reveals that we can take as generators the three elements

$$-\eta, -\psi, \phi.$$

We now need to determine which elements in this subgroup of order 8 give rise to quartics which are locally soluble everywhere. The set of all such elements will form a subgroup which is the 2-Selmer group.

We find that both $-\psi$ and $-\eta\phi$ give quartics which are locally soluble everywhere. But $-\eta$ does not give such a quartic. So the 2-Selmer group has rank two and is given by

$$S_2 = \langle -\psi \rangle \times \langle -\eta\phi \rangle.$$

So we now know that the order of $E(\mathbb{Q})/2E(\mathbb{Q})$ is at most 4. We have three possibilities:

- $|E(\mathbb{Q})/2E(\mathbb{Q})| = 1$ and $|Ш_2| = 4$.
- $|E(\mathbb{Q})/2E(\mathbb{Q})| = 2$ and $|Ш_2| = 2$. This case, however, is not allowed on conjectural grounds as the order of Ш is believed to be always finite and a square.
- $|E(\mathbb{Q})/2E(\mathbb{Q})| = 4$ and $|Ш_2| = 1$.

In general there is no known procedure which is guaranteed to determine which of the above cases we are in. It is for this reason that the weak Mordell–Weil theorem is said to have an ineffective proof.

**XII.2.2. Weak Mordell–Weil theorem: the indirect method.** The only problem with the above method is that one is required to work in the algebra $L$. While using modern computer packages this may now be no problem there is a method which bypasses number fields altogether. This was originally developed by Birch and Swinnerton-Dyer [**15**] in the 1950s. Although this method should be asymptotically much slower than the direct method described above it is rather fast in practice for curves with small discriminant. The method is also explained in [**40**] and [**39**]. Our exposition shall follow this last reference in which the method is explained in terms of classical invariant theory.

The idea is to jump directly to equation (XII.4) and hence bypass the number fields and the first criterion. We therefore do not have the group structure of $K(S,2)$ available to us directly. We are, in some sense, working with the Selmer group only in an indirect manner. We will need to put our elliptic curve into the form

$$E : Y^2 = X^3 - 27IX - 27J$$

where $I$ and $J$ are integers. It is then a fact that the two classical invariants attached to the form $G(m,n)$ in equation (XII.4) (also called $I$ and $J$) are the same as the $I$ and $J$ in the above equation for the elliptic curve (well almost, up to multiplication of $I$ by a fourth power and $J$ by a sixth power).

We shall leave elliptic curves for the moment and quickly go over the classical invariant theory of a binary quartic form. For basic references on this see [**48**], [**100**] or [**161**]. Let $G(X,Y)$ be the binary quartic form

$$G(X,Y) = aX^4 + bX^3Y + cX^2Y^2 + dXY^3 + eY^4,$$

with $a, b, c, d, e \in \mathbb{Q}$. We are interested in such forms up to a modified form of $GL_2(\mathbb{Q})$ equivalence. Two such forms $G$ and $G^*$ will be called equivalent if

$$G(X,Y) = \lambda^2 G^*(\alpha X + \beta Y, \gamma X + \delta Y)$$

where $\lambda \in \mathbb{Q}^*$ and

$$\begin{pmatrix} \alpha & \beta \\ \gamma & \delta \end{pmatrix} \in GL_2(\mathbb{Q}).$$

Classical invariant theory deals with the equivalence relation where the matrix lies in $GL_2(\mathbb{C})$, so it is not surprising we look to invariant theory to help us. There are two basic invariants of a binary quartic form which are

$$I = 12ae - 3bd + c^2$$

of weight 4 and

$$J = 72ace + 9bcd - 27ad^2 - 27eb^2 - 2c^3$$

of weight 6. In addition we shall be interested in the 'seminvariants'. There are (apart from $I$ and $J$) three basic seminvariants,

$$a \ , \ p = 3b^2 - 8ac \ , \ r = b^3 + 8a^2d - 4abc,$$

of weights $0, 2$ and $3$ respectively. They are not, however, algebraically independent as they are linked by the *syzygy*

$$p^3 - 48a^2pI - 64a^3J = 27r^2. \tag{XII.5}$$

Using this syzygy we can describe a rough method for determining a set of binary quartics up to our equivalence relation with given invariants. Remember we are only interested in those binary quartic forms which give rise to an equation such as (XII.4) which is locally soluble everywhere.

We first need a link from the binary quartic forms to the elliptic curve, apart from that in the direct method. For this we need to introduce another two objects from the classical invariant theory of binary quartic forms, namely the two fundamental covariants

$$\begin{aligned} g_4(X,Y) &= -8X^4ac + 3X^4b^2 - 24YaX^3d + 4YbX^3c - 6X^2Y^2bd \\ &\quad +4X^2Y^2c^2 - 48X^2Y^2ae + 4XY^3cd - 24XY^3eb + 3Y^4d^2 \\ &\quad -8Y^4ec, \\ g_6(X,Y) &= -b^3X^6 + 4aX^6bc - 8a^2X^6d - 4YaX^5bd + 8YaX^5c^2 \\ &\quad -2Yb^2X^5c - 32Ya^2X^5e + 20Y^2aX^4cd - 5Y^2b^2X^4d \\ &\quad -40Y^2aX^4eb + 20Y^3aX^3d^2 - 20Y^3b^2X^3e + 5Y^4bX^2d^2 \\ &\quad -20Y^4cX^2eb + 40Y^4dX^2ea + 2Y^5cXd^2 + 32Y^5e^2aX \\ &\quad +4Y^5dXeb - 8Y^5ec^2X + Y^6d^3 + 8Y^6e^2b - 4Y^6edc. \end{aligned}$$

The above syzygy of the seminvariants can be extended to give a syzygy between the covariants:

$$27g_6^2 = g_4^3 - 48IG^2g_4 - 64JG^3. \tag{XII.6}$$

It is this syzygy which provides the link we need. Every binary quartic form, $G(X,Y)$, gives rise to a curve of genus one:

$$C : q^2 = G(m,n). \tag{XII.7}$$

Each such curve is a '2-covering' of $E$ in the sense that there is a diagram

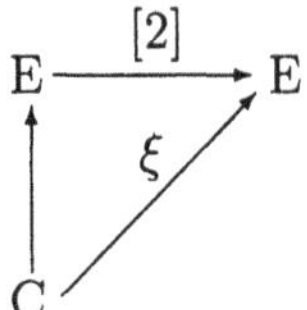


where

$$\xi : \begin{cases} C & \longrightarrow & E \\ (m:q:n) & \longmapsto & (6qng_4(m,n) : 27g_6(m,n) : (2qn)^3). \end{cases}$$

That such a $\xi$ exists follows from the covariant syzygy (XII.6). It can then be shown that the image of $C(\mathbb{Q})$ under $\xi$, if it is non-empty, is a coset of $2E(\mathbb{Q})$ in $E(\mathbb{Q})$ and that all such cosets arise in this way. Hence to determine $E/2E(\mathbb{Q})$ we need to determine all such curves $C$ up to $GL_2(\mathbb{Q})$ equivalence of the associated binary quartic form.

We then need to determine which of the quartics (XII.7) have global solutions. Clearly this leaves us in the same situation as that described earlier. We can determine the 2-Selmer group, $S_2$, but maybe not $E/2E(\mathbb{Q})$. To sum up we have the following:

---

**Birch–Swinnerton–Dyer algorithm for determining $S_2$**

---

DESCRIPTION: Determines $S_2$ as a set of curves of the form $C$.
INPUT: An elliptic curve, $E$, defined over $\mathbb{Q}$.
OUTPUT: A list of curves of the form $C$.

1. Put $E$ in the form $Y^2 = X^3 - 27IX - 27J$ with integral $I$ and $J$.
2. By a change of variable reduce to considering equivalence of binary quartic forms with integer coefficients and invariants coming from a computable finite set.
3. Determine the possible values for the seminvariants of a 'reduced form' in each equivalence class.
4. Given the values of $I, J, a, p, r$ determine values for $b, c, d, e$ from the above formulae.
5. Check all the resulting forms for equivalence under the definition above.
6. Remove from the list all forms such that $q^2 = G(m,n)$ is not locally soluble everywhere.
7. If for all remaining binary quartic forms the curve $C$ has a global solution then we can determine $E(\mathbb{Q})/2E(\mathbb{Q})$ using the

```
map ξ.  Otherwise we have only determined S_2.
```

---

We shall now comment on each step.

Step (2) is achieved using lemma's 3, 4 and 5 of [**15**], as we are only interested in those $G$ such that $q^2 = G(m,n)$ is locally soluble everywhere.

Step (3) is obtained by giving search regions on $a$ and $p$ using a notion of reduction linked to that used by Hermite [**98**] and Julia [**104**]. The details of these search regions can be found in [**15**] and [**40**], but note in these the seminvariant $p$ is labelled $-H$. Given search regions on $a$ and $p$ we can determine possible values of $a, p$ and $r$ by sieving the left hand side of the above syzygy (XII.5) to be 27 times an integral square.

Step (4) is trivial.

Step (5) in [**15**] and [**40**] was achieved by using floating point estimates of certain cross–ratios of roots. This is rather prone to possible floating point errors. In [**39**] a new method is given which depends only on integer arithmetic.

The indirect method above is the one that is implemented in the program **mwrank** by J. Cremona, available by anonymous **ftp** from

**ftp://euclid.ex.ac.uk/pub/cremona/progs/**

## XII.3. The Mordell–Weil theorem

From the previous section we know that $E(\mathbb{Q})/2E(\mathbb{Q})$ is finite and we can, if lucky, determine an explicit set of generators of $E(\mathbb{Q})/2E(\mathbb{Q})$. We can now deduce the full Mordell–Weil theorem using the method of 'infinite descent':

LEMMA XII.8 (infinite descent). *Let $B > 0$ be such that*

$$S = \{P \in E(K) : \hat{h}(P) \leq B\}$$

*contains a complete set of coset representatives for $mE(\mathbb{Q})$ in $E(\mathbb{Q})$, where $m \geq 2$. Then $S$ generates $E(\mathbb{Q})$.*

PROOF. Let $A$ be the subgroup of $E(\mathbb{Q})$ generated by the points in $S$. We suppose that $A$ is not equal to $E(\mathbb{Q})$ and try to deduce a contradiction. Let $Q$ denote a point in $\in E(\mathbb{Q}) - A$ with $\hat{h}(Q)$ minimal, such a $Q$ exists as $\hat{h}$ takes a discrete set of values. By our assumption, we know that there is a $P \in S$ and an $R$ such that $Q = P + mR$. But $R \notin A$, which means $\hat{h}(R) \geq \hat{h}(Q)$ by our choice of $Q$.

Now using the fact that $\hat{h}$ is quadratic and non-negative we obtain

$$\begin{aligned}\hat{h}(P) &= \frac{1}{2}\left(\hat{h}(Q+P)+\hat{h}(Q-P)\right)-\hat{h}(Q),\\ &\geq \frac{1}{2}\hat{h}(mR)-\hat{h}(Q),\\ &= \frac{m^2}{2}\hat{h}(R)-\hat{h}(Q),\\ &\geq \hat{h}(Q)>B,\end{aligned}$$

which contradicts the fact that $P$ was chosen from a set consisting of elements of height less than or equal to $B$. □

Hence if we can find a set of representatives for $E(K)/mE(K)$ we know that a set of generators is contained in the set obtained by enumerating all points with Neron–Tate height less than the maximum Neron–Tate height of the representatives for $E(K)/mE(K)$. But this is easy (at least if $K=\mathbb{Q}$) as we can then bound the naive height of a set of generators and we know, by adapting the sieving method in Section III.4, how to find all rational points on an elliptic curve with bounded $x$-coordinate.

As an example of this we take the curve

$$E: Y^2 = X^3 - 2$$

which we considered earlier. The direct proof of the weak Mordell–Weil theorem led us to determine that we could take the point $P_1 = (3,5)$ as a representative of the only non-trivial element of $E(\mathbb{Q})/2E(\mathbb{Q})$. Silverman's bound on the difference between the naive and the canonical height for this curve gives

$$h(P) \leq \hat{h}(P) + 4.355119.$$

As $\hat{h}(P_1) = 1.349576$ the above lemma tells us that generators for $E(\mathbb{Q})$ are contained in the set of all points with naive height less than $4.355 + 1.349 = 5.704$. So we need to determine all rational points on $E$ with an $X$-coordinate which has numerator and denominator bounded by $\exp(5.704) = 300.65$.

From all these points we need only determine which ones are independent. However, life is slightly easier than this. Suppose $P_1$ does not generate the free part of $E(\mathbb{Q})$. Then it generates a sublattice of full rank with index at least 3. Hence the generator must have canonical height less than $1.349/9 = 0.149$ and naive height less than 4.505. So we are now looking for $X$-coordinates with numerator and denominator bounded by 90. But there are no points with naive height less than 4.505 and canonical height less than 0.149. Hence $E(\mathbb{Q})$ is generated by the point $P_1$.

Another way of deducing $E(\mathbb{Q})$ given $E(\mathbb{Q})/2E(\mathbb{Q})$, is by 'lattice enlargement'. This technique was developed by Siksek [**169**] and it has proved remarkably successful in practice. Suppose we know, by computing $E(\mathbb{Q})/2E(\mathbb{Q})$ for instance, the rank of $E(\mathbb{Q})$ (say $r$) and $r$ independent points on $E(\mathbb{Q})$ of infinite order $P_1, \ldots, P_r$. These $r$ points generate a sublattice of finite index in the free part of $E(\mathbb{Q})$. What we want to do is to bound this index and then enlarge the sublattice until it fills up the whole group.

We first compute a $\lambda$ for which we know that $E(\mathbb{Q})$ contains no points of infinite order with canonical height less than $\lambda$. This may seem similar to the step in the infinite descent above but we can probably use a $\lambda$ much less than the bound $B$ above. Then Hermite's theorem, Theorem V.6, gives us that the index, $n$, of our sublattice in the full lattice is bounded above by

$$I = \sqrt{R(P_1, \ldots, P_r)\gamma_r^r \lambda^{-r}}$$

where $\gamma_r^r$ are Hermite's constants and $R(P_1, \ldots, P_r)$ is the determinant of the height pairing matrix of $P_1, \ldots, P_r$.

We therefore need to detect for each prime, $p$, less than $I$ whether $n$ is divisible by $p$, i.e. can we enlarge the lattice by dividing a point in our sublattice by $p$. So we need to determine if we can solve the equation

$$[p]Q = a_1 P_1 + \cdots + a_r P_r$$

with $a_i \in \{0, \ldots, p-1\}$ and $Q \in E(\mathbb{Q})$. This gives a finite number of cases to check. The number of possible cases can be drastically reduced using a sieving technique which is explained in [**169**].

If the upper bound, $I$, is either less than 1.99 or we deduce that $n$ is not divisible by any prime less than $I$ then we have determined the full Mordell–Weil group. If our $r$ original points had arisen from one of the methods for determining $E(\mathbb{Q})/2E(\mathbb{Q})$ then we know that $n$ can never be divisible by 2, and hence we need only reduce $I$ to a number less than 2.99.

## XII.4. A conditional algorithm

We finally mention another method of computing $E(\mathbb{Q})$ called the conditional algorithm. Those of you who are not experts on elliptic curves can skip this last bit if you so desire.

Firstly make a change of variable so that the elliptic curve is a minimal model. Then for every prime number, $p$, define the quantities $f_p, a_p$ and a polynomial $L_p(T)$ by considering the following cases.

1. **$p$ does not divide $\Delta$.** In this case we say $E$ has good reduction modulo $p$. The group law on $E$ reduces to a well defined group law on $E(\mathbb{F}_p)$. We set

$$f_p = 0 \ , \ a_p = p + 1 - |E(\mathbb{F}_p)| \ , \ L_p(T) = 1 - a_p T + pT^2.$$

2. **$p$ divides $\Delta$ and $c_4$.** We say that $E$ has additive reduction modulo $p$. The non-singular part of the reduction forms a group isomorphic to an additive group, $\mathbb{G}_a$. The 'curve' modulo $p$ has a cusp. We set

$$f_p = 2 + \delta_p \ , \ a_p = 0 \ , \ L_p(T) = 1,$$

where $\delta_p$ is related to the orders of some cohomology groups which define the Swan conductor of a Galois representation. We shall not dwell on this here, except to note that $\delta_p = 0$ if $p > 3$.

3. $p$ **divides** $\Delta$ **but not** $c_4$**.** We say that $E$ has multiplicative reduction modulo $p$. Its 'curve' modulo $p$ looks like a node and the group of non-singular points is isomorphic to a multiplicative group, $\mathbb{G}_m$. We say the curve has split multiplicative reduction if the tangents at the singular point are defined over $\mathbb{F}_p$; otherwise we say it is non-split. We set
$$f_p = 1 \ , \ a_p = \pm 1 \ , \ L_p(T) = 1 + a_p T,$$
where we have $a_p = 1$ in the non-split case.

The conductor and $L$-series of an elliptic curve are then defined to be
$$N_E = \prod_p p^{f_p} \text{ and } L_E(s) = \prod_p L_p(p^{-s})^{-1}.$$

Expanding the above product for the $L$-series we can express the $L$-series as a zeta-function;
$$L_E(s) = \sum_{n \geq 1} \frac{a_n}{n^s},$$
which converges only for all $\Re(s) \geq 3/2$. The $L$–series is believed to satisfy the following conjecture:

CONJECTURE XII.9. *The L-series of an elliptic curve has an analytic continuation to the entire complex plane.*

In fact we have the following more precise formulation:

CONJECTURE XII.10 (Shimura–Taniyama–Weil). *Set*
$$\zeta_E(s) = N_E^{s/2} (2\pi)^{-s} \Gamma(s) L_E(s),$$
*then* $\zeta_E(s)$ *has an analytic continuation to the entire complex plane and*
$$\zeta_E(s) = w \zeta_E(2 - s)$$
*for* $w = \pm 1$.

This last conjecture for semi-stable elliptic curves which Wiles [**217**] proves en route to solving Fermat's last theorem. Indeed he proves a much stronger conjecture. Diamond [**47**] has shown that the conjecture holds for any curve which does not have additive reduction at the primes 2 or 3. Conjecturally the $L$-series should tell us everything we need to know about the elliptic curve. As a start we have:

CONJECTURE XII.11 (weak Birch–Swinnerton–Dyer). $L_E(s)$ *has a zero at* $s = 1$ *of order equal to the rank of* $E(\mathbb{Q})$.

So if we can compute the order of vanishing of the $L$-series of $E$ at the point $s = 1$ then we can compute the rank of the group $E(\mathbb{Q})$. This is the idea behind the conditional algorithm [**79**]. It is conditional on all the above conjectures.

Given Conjecture XII.10 we can compute numerical estimates for the value of the $L$-function and its derivatives at $s = 1$, see [**40**] or [**32**]. These are, however, only numerical values, and with any floating point numerical computation there is the question whether we can detect zero. To solve this problem we need to introduce a refined version of the Birch–Swinnerton–Dyer conjecture:

CONJECTURE XII.12 (strong Birch–Swinnerton–Dyer). *There is a computable constant, $C$ (usually called the 'fudge factor'), such that*

$$\lim_{s\to 1}(s-1)^{-r}L_E(s) = C\frac{R(E)|\text{Ш}|}{|E_{tors}|^2},$$

*where the quantity $|\text{Ш}|$ is always a positive integer.*

Assume we compute a numerical value for $\lim_{s\to 1}(s-1)^{-r}L_E(s)$ which is so close to zero we suspect it is actually zero. We assume that it is not zero and try to produce a contradiction. We have a small numerical value for $\alpha$ such that

$$\alpha \geq |\lim_{s\to 1}(s-1)^{-r}L_E(s)| \geq \epsilon,$$

where $\epsilon$ is some positive constant. So $E$ has rank $r$ and $R(E) \leq \alpha T^2/C$, where $T$ denotes the order of the torsion group of $E(\mathbb{Q})$. However, by Hermite's theorem, Theorem V.6, there should exist a point, $P$, on $E$ such that

$$\hat{h}(P)^r \leq \mu_r R(E) \leq \mu_r \alpha T^2/C \leq \mu_r \alpha T^2.$$

So there should be points on $E$ of very small height. If we cannot find such a point then we must conclude that the value of $\lim_{s\to 1}(s-1)^{-r}L_E(s)$ under consideration is actually zero.

Using the above technique we can determine the exact order of vanishing, $r$, of the $L$-series at $s = 1$. We now search for $r$ independent points on the curve. Once we have found $r$ such points we know a sublattice of finite index in $E(\mathbb{Q})$ and hence we can apply the lattice enlarging procedure to determine the full Mordell–Weil group.

## XII.5. Exercises

1). Derive the formulae for the group law on an elliptic curve.

2). Work out the details for the direct method of computing $E(\mathbb{Q})/2E(\mathbb{Q})$ for the case when $E(\mathbb{Q})$ has three non-trivial points of order 2. In other words the algebra $L$ decomposes into the sum of three copies of $\mathbb{Q}$.

3). Fill in all the details for the example $Y^2 = X^3 - 141$. In particular show that the elements given in the text do indeed generate $K(S,2)$ and that the Selmer group is what we claim it to be.

4). Compute the 2-Selmer groups of the following elliptic curves:

1. $Y^2 = X^3 - 9X$.
2. $Y^2 = X^3 - 6X^2 + 11X - 6$.
3. $Y^2 = X^3 + 17X$.
4. $Y^2 = X^3 + X + 1$.
5. $Y^2 = X^3 + 6X + 432$.
6. $Y^2 + Y = X^3 - X^2 - 929X - 10595$.

For each of these curves determine generators for $E(\mathbb{Q})/2E(\mathbb{Q})$. If this is not possible, determine why it is difficult and try to find a method of making it possible

5). Show that the map $\Theta$ which was used in the direct method for computing $E(\mathbb{Q})/2E(\mathbb{Q})$ is indeed a group homomorphism with kernel $E(\mathbb{Q})$.

## CHAPTER XIII

# Integral points on elliptic curves

In previous chapters we have seen how to solve the problem of finding all integral points on an elliptic curve. The methods used either a reduction to a finite set of Thue equations or reduction to a finite set of $S$-unit equations. These methods had numerous drawbacks in that they involved using expensive computations in number fields and they ignored much of the beauty of elliptic curves.

In this chapter we present a much better method which uses a lot of the underlying structure of an elliptic curve. The new method is based on the method of elliptic logarithms. The idea behind this method can be found in a paper by Lang from 1964 [**110**]. It is also explained in [**111**] or [**172**], and an outline of the method was also given in [**219**]. However, it was not until David [**44**] gave an explicit transcendence result for elliptic logarithms that it became a general method. This method is now the standard one, which is apparent from looking at the relevant literature [**191**], [**77**], [**179**], [**192**], [**189**], [**201**] and [**185**].

There is one drawback with the new method in that we need to be able to compute the Mordell–Weil group. In other words to find all integral points we shall need an explicit description of the set of all rational points on the curve. As we saw in the last chapter this may be a major problem as our algorithms for determining the Mordell–Weil group may not work for the example we are interested in.

One minor problem is that there are no efficient techniques for the final search where one tries to locate all the small solutions. At present no generalization of the method of sieving or an analogue of the method of Bilu and Hanrot is known. This means we need to restrict attention to elliptic curves with 'small' rank, say less than 8. In practice this is no problem as 'most' elliptic curves do have very small ranks.

## XIII.1. Elliptic logarithms

Let $E$ be an elliptic curve given by

$$Y^2 + a_1XY + a_3Y = X^3 + a_2X^2 + a_4X + a_6$$

with $a_i \in \mathbb{Z}$. As mentioned before, any elliptic curve is isomorphic to a curve of the form

$$E' : Y^2 = 4X^3 - g_2X - g_3 = 4(X - e_1)(X - e_2)(X - e_3) = f(X),$$

where $g_2,\ g_3 \in \mathbb{Z}$. We let $\wp(z)$ be the solution to the associated differential equation

$$\wp'(z)^2 = 4\wp(z) - g_2\wp(z) - g_3.$$

Such a function, $\wp(z)$, is called the Weierstrass elliptic function with parameters $g_2$ and $g_3$. This is a doubly periodic function with periods $\omega_1$ and $\omega_2$. The periods, $\omega_1$ and $\omega_2$, form a basis of a lattice in the complex plane. The standard (extensive) theory of elliptic functions, see [**215**], tells us that a basis for this lattice is given by the integrals

$$\begin{aligned}\omega_1 &= 2\int_{e_1}^{e_2} \frac{dX}{\sqrt{4(X-e_1)(X-e_2)(X-e_3)}},\\ \omega_2 &= 2\int_{e_3}^{e_2} \frac{dX}{\sqrt{4(X-e_1)(X-e_2)(X-e_3)}}.\end{aligned}$$

We shall show later that we can always choose $\omega_1 \in \mathbb{R}$ and $\Im(\omega_1/\omega_2) > 0$, if our curve is defined over the real numbers. In what follows we shall therefore assume that $\omega_1 \in \mathbb{R}$ and $\Im(\omega_1/\omega_2) > 0$. Let $\Lambda = \mathbb{Z}\omega_1 + \mathbb{Z}\omega_2$ be the period lattice; then we have the map

$$\begin{array}{ccc}\mathbb{C}/\Lambda & \xrightarrow{\phi} & E\\ z & \longmapsto & \left\{\begin{array}{cc}(\wp(z)-b_2/12, (\wp'(z)-a_1x-a_3)/2) & z\notin\Lambda\\ \infty & z\in\Lambda\end{array}\right. .\end{array}$$

The 'inverse' of this map, after choosing a fundamental region, we will call the elliptic logarithm, $\psi$. It is given by

$$\psi(P) = \int_{\infty}^{x+b_2/12} \frac{dt}{\sqrt{4t^3 - g_2t - g_3}} \pmod{\Lambda}.$$

It is usual to take the fundamental region to be $\{a\omega_1 + b\omega_2\ :\ a, b \in \mathbb{R}, 0 \leq a, b < 1\}$, in which case we state the $\psi$ takes its principal value. Why should this map be called a 'logarithm'? Well it satisfies the property

$$\psi(P+Q) = \psi(P) + \psi(Q) \pmod{\Lambda}$$

which is rather like the relationship between complex logarithms

$$\log(xy) = \log(x) + \log(y) \pmod{2\pi\sqrt{-1}\mathbb{Z}}.$$

## XIII.2. Elliptic integrals and the *AGM*

For our method to compute integral points we will require a method to compute the periods and the value of elliptic logarithms accurately to a large number of decimal digits. In this section we will review the theory needed; for those readers who just want to implement the algorithms see the pseudocode in the book by Cohen [**32**].

We first need to introduce Gauss's *arithmetic–geometric mean* of two numbers $a$ and $b$, which is usually written as $AGM(a,b)$. To compute the $AGM(a,b)$ we compute the two sequences $(a_n), (b_n)$ such that

$$a_0 = a\ ,\ b_0 = b\ ,\ a_{n+1} = (a_n + b_n)/2\ ,\ b_{n+1} = \sqrt{a_n b_n}.$$

If we start off with two positive real numbers $a$ and $b$ then the two sequences above will converge to the same real number, which is denoted $AGM(a,b)$. Such a limit was discovered by Lagrange before 1785 and then rediscovered in the next decade by Gauss.

If either $a$ or $b$ is not a positive real number then the sequences will converge to one of a countable number of possible limits. This of course depends on us making a choice for the value of the complex square root. The 'correct' choice is that $b_{n+1}$ should satisfy

$$|a_{n+1} - b_{n+1}| \leq |a_{n+1} + b_{n+1}|$$

and if we have equality then we insist that $\Im(b_{n+1}/a_{n+1}) > 0$.

In any case the convergence of the $AGM$ is quadratic in nature, so we do not have to take many terms to obtain a very accurate answer. The proof of convergence in the positive real case is easy; however, the proof for the general case is less obvious. For more details on the $AGM$ you should consult the articles [**37**] and [**38**].

For example suppose we wish to compute the $AGM$ of the numbers $a_0 = 1$ and $b_0 = \sqrt{2}$. We then find

$$\begin{array}{ll} a_1 = 1.20710678118654752440 & b_1 = 1.18920711500272106671 \\ a_2 = 1.19815694809463429555 & b_2 = 1.19812352149312012260 \\ a_3 = 1.19814023479387720908 & b_3 = 1.19814023467730720579 \\ a_4 = 1.19814023473559220744 & b_4 = 1.19814023473559220743 \end{array}$$

where the numbers have been computed to twenty decimal digits accuracy. We can see the quadratic convergence due to the doubling of the accuracy at every stage

$$\begin{aligned} |a_1 - b_1| &\leq 10^{-1}, \\ |a_2 - b_2| &\leq 10^{-4}, \\ |a_3 - b_3| &\leq 10^{-9}, \\ |a_4 - b_4| &\leq 10^{-20}, \end{aligned}$$

To compute our periods we need to link the $AGM$ with the integrals giving the periods. This is done using another integral denoted by

$$I(a,b) = \int_0^{\pi/2} \frac{d\theta}{\sqrt{a^2\cos^2\theta + b^2\sin^2\theta}}.$$

We shall also a little later have need to consider the integral

$$I(a,b,X) = \int_0^{X} \frac{d\theta}{\sqrt{a^2\cos^2\theta + b^2\sin^2\theta}}.$$

First we have the result:

LEMMA XIII.1. *Let $a, b \in \mathbb{C}$ then*

$$2I(a,b)AGM(a,b) = \pi.$$

PROOF. We first notice that it is easy to compute $I(c,c,X) = X/c$. Then consider the integral for $I(a,b)$ given above and perform the substitution

$$\sin\theta_1 = \frac{a_1 \sin 2\theta}{\sqrt{a^2\cos^2\theta + b^2\sin^2\theta}}$$

where (as above) $a_1 = (a+b)/2$ and $b_1 = \sqrt{ab}$. Then we find (after a lot of messy algebra)

$$I(a,b,X) = I(a_1,b_1,X_1)/2$$

where

$$\sin X_1 = \frac{a_1 \sin 2X}{\sqrt{a^2\cos^2 X + b^2\sin^2 X}}.$$

In particular we have

$$I(a,b) = I(a,b,\pi/2) = I(a_1,b_1,\pi)/2 = I(a_1,b_1,\pi/2) = I(a_1,b_1).$$

We can clearly repeat this substitution over and over again so we find

$$\begin{aligned} I(a,b) = I(a_1,b_1) = I(a_2,b_2) = \cdots &= I(AGM(a,b), AGM(a,b)) \\ &= \frac{\pi}{2AGM(a,b)}. \end{aligned}$$

From which the result follows. □

The link the other way is given by:

LEMMA XIII.2.

$$\omega_1 = 2\int_{e_1}^{e_2} \frac{dX}{\sqrt{4(X-e_1)(X-e_2)(X-e_3)}} = 2I(\sqrt{e_3-e_1}, \sqrt{e_3-e_2}).$$

$$\omega_2 = 2\int_{e_3}^{e_2} \frac{dX}{\sqrt{4(X-e_1)(X-e_2)(X-e_3)}} = 2\sqrt{-1}I(\sqrt{e_3-e_1}, \sqrt{e_2-e_1}).$$

PROOF. For the first result make the substitution $X = e_1 + (e_2-e_1)\sin^2\theta$, whilst for the second make the substitution $X = e_3 + (e_2-e_3)\sin^2\theta$. □

Then, putting this altogether, we see that

$$\begin{aligned} \lambda_1 &= \frac{\pi}{AGM(\sqrt{e_3-e_1}, \sqrt{e_3-e_2})}, \\ \lambda_2 &= \frac{\pi\sqrt{-1}}{AGM(\sqrt{e_3-e_1}, \sqrt{e_2-e_1})} \end{aligned}$$

will converge to some periods of the functions $\wp$.

If our elliptic curve has coefficients in $\mathbb{R}$ we can make life easier by using the positive real version of the $AGM$ algorithm as much as possible. We have two cases to consider

1. $\Delta > 0$. In this case $e_1, e_2, e_3$ are all real and we can order them so that $e_3 > e_2 > e_1$, in which case (assuming we always take the positive square root of positive real numbers) the two $AGM$'s in the formula for $\lambda_1$ and $\lambda_2$ are both $AGM$'s of positive real numbers. Then clearly we can use the positive real version of the $AGM$ and $\lambda_1$ will be a positive real number and $\lambda_2$ will have a positive imaginary part. Hence $\omega_1 = \lambda_1$ and $\omega_2 = \overline{\lambda_2}$.
2. $\Delta < 0$. Now let $e_3$ denote the single real root. Then setting $z = \sqrt{e_3 - e_1}$ we find $\overline{z} = \sqrt{e_3 - e_2}$. We reorder $e_1$ and $e_2$ so that $\Re(z) > 0$. Then
$$\begin{aligned}\lambda_1 &= \frac{\pi}{AGM(z,\overline{z})} = \frac{\pi}{AGM((z+\overline{z})/2, \sqrt{z\overline{z}})}\\ &= \frac{\pi}{AGM(\Re(z), |z|)}\end{aligned}$$
Hence to compute $\lambda_1$ we only require the positive real version of the $AGM$ and the result will certainly be real and positive. So we can take $\omega_1 = \lambda_1$ and we can then fix $\omega_2$ to satisfy $\Im(\omega_1/\omega_2) > 0$ as required.

So now having a method to compute the periods we need a method to compute the elliptic logarithm of a point. To compute $\psi$ we use the integral $I(a, b, X)$ mentioned above with $a = \sqrt{e_3 - e_1}$ and $b = \sqrt{e_2 - e_1}$ as

$$\psi(P) = \int_\infty^{e_3} + \int_{e_3}^{x+b_2/12} \frac{dt}{\sqrt{4t^3 - g_2 t - g_3}} \pmod{\Lambda} = \frac{\omega_1}{2} + I(a,b,\phi) \pmod{\Lambda}$$

where $x = e_3 - b_2/12 + (e_2 - e_3)\sin^2\phi$. We then compute $(a_1, b_1, \phi_1)$, $(a_2, b_2, \phi_2)$, etc. where

$$a_{i+1} = (a_i + b_i)/2\ ,\ b_{i+1} = \sqrt{a_i b_i}\ ,\ \sin\phi_{i+1} = \frac{a_i \sin 2\phi_i}{\sqrt{a_i^2\cos^2\phi_i + b_i^2\sin^2\phi_i}}.$$

until $|a_n - b_n|$ is very, very small. So we have (approximately) $a_n \approx b_n \approx AGM(a, b)$. Then

$$I(a,b,\phi) = \frac{I(a_1,b_1,\phi_1)}{2} = \frac{I(a_2,b_2,\phi_2)}{4} = \cdots = \frac{I(a_n,b_n,\phi_n)}{2^n} \approx \frac{\phi_n}{a_n 2^n}.$$

Note that although we can compute the $a_n$ very accurately we have a problem computing the $\phi_n$. This will only be as good as our arcsin function, so particular care needs to be taken on how we compute arcsin.

There is another method due to Zagier [**219**] for computing elliptic logarithms which avoids the use of any transcendental functions. In addition Zagier's method allows one to explicitly bound the error term that one has in computing the elliptic logarithm. The method only relies on elementary operations and no evaluation of arcsin, which makes it somewhat easier to implement.

There are two major drawbacks with Zagier's method in that it takes longer to get the accuracy one wants and it only works for points $P$ such that

$P \in E^0(\mathbb{R})$. That it is slow is because it computes the elliptic logarithm one bit at a time via the formula

$$\psi(P) = \omega_1 \sum_{i=0}^{\infty} \frac{a_i}{2^{i+1}},$$

where

$$a_i = \begin{cases} 0 & \text{if } y([2^i]P) > 0, \\ 1 & \text{if } y([2^i]P) < 0. \end{cases}$$

That this works can be seen from the fact that if $P \in E^0(\mathbb{R})$, such that $[2]P \neq 0$, then $\psi(P) \in (\omega_1/2, \omega_1)$ if $y(P) < 0$ and $\psi(P) \in (0, \omega_1/2)$ if $y(P) > 0$.

## XIII.3. Integral points

Unlike before, when we considered the problem of finding integral points on an elliptic curve where we only considered curves of the form

$$Y^2 = X^3 + aX^2 + bX + c,$$

we shall now look for integral points on a general elliptic curve given by an equation in long Weierstrass form

$$Y^2 + a_1XY + a_3Y = X^3 + a_2X^2 + a_4X + a_6$$

where $a_i \in \mathbb{Z}$. We shall assume we know a basis, $P_1, \ldots, P_r$, for the free part of $E(\mathbb{Q})$. We hope the methods in the last chapter have allowed us to compute such a basis, and hence we can write

$$E(\mathbb{Q}) = \langle P_1 \rangle \times \cdots \times \langle P_r \rangle \times \mathrm{Tors}(E).$$

If $P$ denotes our integral point then we may write it as

$$P = p_1P_1 + \cdots + p_rP_r + T \qquad \text{(XIII.1)}$$

where $T$ is some element of $\mathrm{Tors}(E)$ and $p_i \in \mathbb{Z}$. So we need to determine which values the variables $p_i$ can take to make the point $P$ integral. It is a simple matter to determine all integral points on the egg, and therefore we shall assume that $P \in E^0(\mathbb{R})$.

We define $m_i$ for $i = 1, \ldots, r$ by the condition

$$m_i = \begin{cases} 1 & \text{if } P_i \in E^0(\mathbb{R}), \\ 2 & \text{if } P_i \text{ is on the egg.} \end{cases}$$

It is then clear that for $i = 1, \ldots, r$ we have $Q_i = m_iP_i \in E^0(\mathbb{R})$. We then define $q_i \in \mathbb{Z}$ by euclidean division of the $p_i$ in equation (XIII.1):

$$p_i = m_iq_i + r_i \quad \text{with } 0 \leq r_i < m_i.$$

Hence if we set

$$U = r_1P_1 + \cdots + r_rP_r$$

then we can write our integral point $P$ as

$$P = q_1Q_1 + \cdots + q_rQ_r + T + U \qquad \text{(XIII.2)}$$

Now as $P \in E^0(\mathbb{R})$ and $Q_i \in E^0(\mathbb{R})$ we must have $T+U \in E^0(\mathbb{R})$ as well. We shall set $Q_{r+1} = T+U$ and so $Q_{r+1}$ comes from a computable finite set.

We shall put $H = \max|q_i|$. Clearly if we can find a small upper bound on $H$ then we can enumerate all the possible integral points and we shall be done. Set $K = \max|p_i|$ and note that $H \le K$. To bound $H$ we first need to link $H$ to the size of the $x$-coordinate of the integral point $P$; this is accomplished in the next lemma:

LEMMA XIII.3. *With the notation above, if $P$ is an integral point,*

$$\frac{1}{|x(P)|} \le c_1 e^{-c_2 H^2},$$

*for two computable constants $c_1$ and $c_2$.*

PROOF. As $P$ is an integral point we have $h(P) = \log|x(P)|$. From Lemma XII.4 we know $h(P) \ge \hat{h}(P) - c_3$, hence $\log|x(P)| \ge \hat{h}(P) - c_3$. Let $R$ denote the regulator matrix defined earlier, i.e.

$$R = (\langle P_i, P_j \rangle),$$

and then $\hat{h}(P) = \vec{p}\,^t R \vec{p}$. We can compute an orthogonal decomposition of $R$ into $O^t \Lambda O$, where $O$ is an orthogonal matrix and $\Lambda$ is a diagonal matrix of eigenvalues of $R$. We put

$$c_2 = \min_{i=1\ldots r} \Lambda_{i,i}$$

and $\vec{m} = O\vec{p}$, then as $O^t O = I$ we have

$$\begin{aligned}
\hat{h}(P) &= \vec{p}\,^t R \vec{p} = \vec{m}\,^t \Lambda \vec{m} = \sum_{i=1}^{r} \Lambda_{i,i} m_i^2 \\
&\ge c_2 \sum_{i=1}^{r} m_i^2 = c_2 \vec{m}\,^t \vec{m} = c_2 \vec{p}\,^t O^t O \vec{p} = c_2 \vec{p}\,^t \vec{p} \\
&= c_2 \sum_{i=1}^{r} p_i^2 \ge c_2 K^2.
\end{aligned}$$

The result is then immediate on putting $c_1 = \exp(c_3)$, as $K \ge H$. □

Our second link is between the size of the $x$-coordinate of the integral point $P$ and the size of the elliptic logarithm of $P$. As mentioned before, our elliptic curve is isomorphic to a curve of the form

$$E' : Y^2 = 4X^3 - g_2 X - g_3 = f(X).$$

We let $\gamma_1, \gamma_2, \gamma_3$ denote the complex roots of $f(X)$ from which we can compute the constant

$$c_4 = 2 \max_{i=1,2,3} |\gamma_i|.$$

LEMMA XIII.4. *If $P \in E^0(\mathbb{R})$ and $|x(P)+b_2/12| > c_4$ then*

$$|\psi(P)|^2 \leq \frac{c_5}{|x(P)|},$$

*where $c_5 = 8 + |\omega_1^2 b_2|/12$.*

PROOF. We put $X(P) = x(P) + b_2/12$. Then, as we remarked before, the elliptic logarithm is defined by

$$\psi(P) = \int_\infty^{X(P)} \frac{dt}{\sqrt{f(t)}} \quad (\text{mod } \Lambda).$$

The integral is along a path along the real axis in a negative direction. As we are assuming that $|X(P)| > c_4$ then for all $t \geq |X(P)|$ we have

$$|t|/2 \leq |t - \gamma_i| \text{ for } i = 1, 2, 3,$$

and so $|f(t)| > \frac{1}{2}|t|^3$. Now for all $N \in \mathbb{R}_+$ with $N \geq |X(P)| > c_4 > 0$ we have

$$\begin{aligned} \left| \int_N^{|X(P)|} \frac{dt}{\sqrt{f(t)}} \right| &\leq \sqrt{2} \left| \int_N^{|X(P)|} |t|^{-3/2} dt \right| \\ &= 2\sqrt{2}\left( |X(P)|^{-1/2} - N^{-1/2} \right). \end{aligned}$$

Letting $N$ tend to $\infty$ then gives

$$|\psi(P)| \leq 2\sqrt{2}|X(P)|^{-1/2}.$$

Hence

$$|x(P) + b_2/12| = |X(P)| \leq \frac{8}{|\psi(P)|^2}.$$

But then

$$\begin{aligned} |x(P)| &= |x(P) + b_2/12 - b_2/12|, \\ &\leq |x(P) + b_2/12| + |b_2|/12, \\ &\leq \frac{8}{|\psi(P)|^2} + |b_2|/12, \\ &\leq \frac{8 + |\psi(P)|^2 |b_2|/12}{|\psi(P)|^2}, \\ &\leq \frac{c_5}{|\psi(P)|^2}, \end{aligned}$$

this last inequality following because as $P \in E^0(\mathbb{R})$ we have $|\psi(P)| \leq |\omega_1|$. □

Combining these last two lemmas we obtain, for an integral point $P \in E^0(\mathbb{R})$ with $|x(P) + b_2/12| > c_4$,

$$|\psi(P)| \leq \sqrt{c_5 c_1} e^{-c_2 H^2/2} = c_6 e^{-c_7 H^2}. \quad \text{(XIII.3)}$$

But the left hand side of the above inequality can be written as

$$|q_1\psi(Q_1) + \cdots + q_r\psi(Q_r) + \psi(Q_{r+1}) + m\omega_1|$$

where $|m| \leq rK + 2$. Note that this is a linear form in elliptic logarithms and hence, by the result of David, Appendix A.3, it can be bounded from below by

$$-c_8(\log H + c_9)(\log\log H + c_{10})^{r+2}.$$

Comparing this with the function $c_6 e^{-c_7 H^2}$ we can deduce a very large upper bound, $H_0$, on $H$. Then using the reduction techniques of VI.3 on the inequality (XIII.3) we can reduce this large upper bound on $H$ to something of the order of $\sqrt{\log H_0}$. This should be small enough to enumerate all the possibilities for integral points.

There are a few points to note:

1. As every entry in the linear form in elliptic logarithms is an elliptic logarithm of a point on $E^0(\mathbb{R})$ then the elliptic logarithm is real and we can use Zagier's method to compute the elliptic logarithm if we want.
2. Because we obtain roughly the square root of the logarithm of the original large bound in the reduction process this method is very good at producing final upper bounds which are very small indeed.
3. The main problem is that we need a basis for the Mordell–Weil group of $E$ before we can even apply the method.

Note the similarities of this method with the algorithm for solving an $S$-unit equation. In both cases one has a finitely generated abelian group and logarithm maps for which we know effective lower bounds on linear forms in the logarithms of the generators. We translate our diophantine problem into determining when such a linear form is very small and then use our effective lower bound to say that the solutions to our diophantine problem must be finite in number.

You may ask why the method using elliptic logarithms is better than the method using Thue equations. After all to apply the method using elliptic logarithms we need to know a basis for the Mordell–Weil group. That the method of elliptic logarithms is better is best illustrated with an example.

For the 28th International Mathematical Olympiad, Juan Ochoa Melida proposed that one question should be to determine the integral points on the curve

$$3Y^2 = 2X^3 + 385X^2 + 256X - 58195.$$

This is now known as the Ochoa curve. Guy [**85**] wondered how the proposal intended the contestants to solve this problem. This problem is equivalent to finding the integral points on the following elliptic curve in short Weierstrass form:

$$E : y^2 = x^3 - 440067x + 106074110.$$

It turns out, see [**192**], that it is virtually impossible to determine the integral points on this curve using the method which goes via Thue equations.

However, the Mordell–Weil group of $E$ has rank four, a basis being given by

$$(247, 3528)\ ,\ (499, 3276)\ ,\ (751, 14112)\ ,\ (-761, 504).$$

It is then rather easy to implement the previous method. Stroeker and de Weger [**192**] implement the method and show that the elliptic curve $E$ has 46 integral points, the largest of which has $x$-coordinate given by 1657691.

Why should this be the case? Remember that the method using Thue equations leads to the need to solve quartic Thue equations. Hence we need to deduce fundamental units etc. in lots of quartic number fields. However, we do not require any number field data to compute the Mordell–Weil group, if we use the indirect method. If we use the direct method we only need the data for a single cubic number field. It is this fact which means that the method using elliptic logarithms is more efficient.

The most extensive computations with the method of elliptic logarithms have been on Mordell's equation,

$$y^2 = x^3 + k,$$

where $k \in \mathbb{Z}$. Computational resolution of this equation goes back to the 1960s with work of Ljunggren [**123**] and Coghlan and Stephens [**31**]. In recent years the number of values for $k$ for which we know all the integral points has increased dramatically, see [**78**].

## XIII.4. Integral points on the curve $Y^2 = X^3 - 2$

In this section we look in detail at the curve $E : Y^2 = X^3 - 2$. As we proved in the last chapter, $E(\mathbb{Q})$ has rank one and trivial torsion group. As a generator of the group $E(\mathbb{Q})$ we can take the point $P_1 = (3, 5)$. We would like to determine all other integral points on this curve. To do this we apply the method just discussed.

We find, using the $AGM$ mentioned earlier (or using the package **PARI**), that

$$\begin{aligned} \psi(P_1) &\approx 1.58318127, \\ \omega_1 &\approx 2.16368175, \\ \omega_2 &\approx 1.081840875 - 1.873803360\sqrt{-1}. \end{aligned}$$

If $P$ is an integral point on $E(\mathbb{Q})$ we write $P = nP_1$. Our task is then to bound $n$ and enumerate all possible values of $n$ up to the bound we have found.

From Lemma XIII.3 we find that for such a point $P$ on our curve we have the inequality

$$\frac{1}{|x(P)|} \leq 94.54917e^{-0.74097n^2}.$$

We also find from Lemma XIII.4 that if $|x| > 2.5$ then

$$|\psi(P)|^2 \leq \frac{8}{|x(P)|}.$$

In other words, putting these last two inequalities together,

$$|\psi(P)| \leq 27.50261e^{-0.37049n^2}. \qquad \text{(XIII.4)}$$

But $|\psi(P)|$ is the linear form in elliptic logarithms given by

$$|\psi(P)| = |n\psi(P_1) + m\omega_1|$$

where $|m| \leq |n|$. Using David's result we can find the following lower bound on $|\psi(P)|$, assuming $|n| \geq 53252$:

$$-c_8(\log|n| + c_9)(\log\log|n| + c_{10})^3 < \log|\psi(P)|,$$

where $c_8 = 1.4 \cdot 10^{39}$, $c_9 = 1$ and $c_{10} = 3.07944$. This allows us to deduce that

$$|n| \leq 10^{22}.$$

This is clearly far too large to enumerate all possible values of $n$ so we reduce this upper bound using inequality (XIII.4). We find that a reduced basis of the lattice, $\mathcal{L}$, generated by the columns of the matrix

$$\begin{pmatrix} 1 & 0 \\ [10^{45}\psi(P_1)] & [10^{45}\omega_1] \end{pmatrix}$$

is given by the columns of the matrix

$$\begin{pmatrix} -28559795545411285398829 & 24948051935918593148497 \\ -28602825742743919175758 & -50774066801844956867765 \end{pmatrix}.$$

This allows us to compute that $\ell(\mathcal{L}, \vec{0})$ is bounded from below by $10^{22}$. Then application of our standard lemma from Section VI.3 gives us that

$$|n| \leq 12.$$

It is a much easier task to enumerate all points below this new bound. If we do this we find that the only integral points on $E$ are given by $n = \pm 1$, which corresponds to the points $(3, \pm 5)$.

## XIII.5. $S$–integral points

By an $S$-integral point on an elliptic curve we mean a rational point which has an $x$-coördinate with denominator divisible only by the primes in a fixed finite set $S$. Our above method for integral points can easily be generalized to $S$-integral points, again assuming a basis of the Mordell–Weil group is known. This is due to the existence of a $p$-adic elliptic logarithm map. Alas at the time of writing, the theory of lower bounds for linear forms in $p$-adic elliptic logarithms is not at such an advanced state as the theory of ordinary elliptic logarithms. The method below will therefore, at present, not guarantee the existence of all $S$-integral points but it will find all of them up to a very large given bound.

Let $S$ denote a finite set of places of $\mathbb{Q}$ including the infinite place. We are looking for all points on an elliptic curve $E(\mathbb{Q})$ such that the $x$-coordinate has denominator supported on $S$. As before we deduce an upper bound, in terms of $K$, on the inverse of the absolute value of the $x$-coordinate.

LEMMA XIII.5. *There are positive computable constants $c_{11}$ and $c_{12}$ such that if $P$ is an $S$-integral point then for some $v \in S$ we have*

$$\frac{1}{|x(P)|_v} \leq c_{11} e^{-c_{12} K^2}.$$

PROOF. Exercise. □

We now just have to work out what the above lemma implies for each possible $v \in S$. Now if $v$ is the infinite place then we can use verbatim the method for integral points above. We therefore only have to consider what to do when $v$ is a finite place, corresponding to a prime number $p$.

In what follows it will be convenient to consider $E(\mathbb{Q})$ embedded in $E(\mathbb{Q}_p)$. Standard elliptic curve theory tells us that

$$|x([m]P)|_p \geq |x(P)|_p.$$

Hence we can certainly multiply $P$ by any integer $m$ and still have the inequality of the above lemma. We choose an $m$ such that $[m]P$ will always lie in the 'kernel of reduction modulo $p$'. That is the point $[m]P$ when considered as a point on the reduced curve modulo $p$ is actually the point at infinity, i.e. the zero of the group law. The set of points of $E(\mathbb{Q}_p)$ which reduce to zero modulo $p$ is denoted $E_1(\mathbb{Q}_p)$ which is a group. Hence we choose an $m$ such that $[m]P \in E_1(\mathbb{Q}_p)$ for all $P \in E(\mathbb{Q})$, any multiple of $|E(\mathbb{F}_p)|$ will do. We may as well also choose $m$ to be a number which also kills off all of the torsion in $E(\mathbb{Q})$ as well.

Now $E_1(\mathbb{Q}_p)$ is isomorphic to another group, the one parameter formal group associated to $E$, see [**172**, p. 175]. Explicitly we have the isomorphism

$$\psi_p : \begin{cases} \hat{E}(p\mathbb{Z}_p) & \longrightarrow & E_1(\mathbb{Q}_p) \\ z & \longmapsto & \begin{cases} \infty & \text{if } z = 0 \\ \left(\frac{z}{w(z)}, \frac{-1}{w(z)}\right) & \text{otherwise, i.e. } z = -x/y \end{cases} \end{cases}$$

where $w(z)$ is a power series in $z$, which is the formal power series solution to the equation

$$w = z^3 + a_1 z w + a_2 z^2 w + a_3 w^2 + a_4 z w^2 + a_6 w^3.$$

Such a solution can be computed to any desired number of terms using the Newton–Raphson iteration. Every point $P \in E_1(\mathbb{Q}_p)$ is the image of an element $z_P$ in the formal group, $\hat{E}(p\mathbb{Z}_p)$. We let $\omega(z)$ denote the invariant differential on $\hat{E}(p\mathbb{Z}_p)$. From the power series for $w(z)$ we can compute the

Laurent series for $x(z)$, $y(z)$ and $\omega(z)$, again see [**172**]. They turn out to have the first few terms given by

$$\begin{aligned} x(z) &= \frac{z}{w(z)} = \frac{1}{z^2} - \frac{a_1}{z} - a_2 - a_3 z - (a_4 + a_1 a_3) z^2 - \cdots \\ y(z) &= \frac{-1}{w(z)} = \frac{-1}{z^3} + \frac{a_1}{z^2} + \frac{a_2}{z} + a_3 + (a_4 + a_1 a_3) z + \cdots \\ \omega(z) &= \frac{dx(z)}{2y(z) + a_1 x(z) + a_3} \\ &= \left(1 + a_1 z + (a_1^2 + a_2) z^2 + (a_1^3 + 2a_1 a_2 + a_3) z^3 + \cdots\right) dz \\ &= (1 + d_1 z + d_2 z^2 + d_3 z^3 + \cdots) dz. \end{aligned}$$

Then for a point on $E_1(\mathbb{Q}_p)$ we define the $p$-adic elliptic logarithm to be the map

$$\psi_p : \begin{cases} E_1(\mathbb{Q}_p) & \longrightarrow \quad \hat{\mathbb{G}}_a \\ P & \longmapsto \quad \int \omega(z_p) = z_p + \frac{d_1 z_p^2}{2} + \frac{d_2 z_p^3}{3} + \cdots \end{cases}$$

This has all the properties we require, in that it is a homomorphism since

$$\psi_p(P + Q) = \psi_p(P) + \psi_p(Q),$$

and it satisfies

$$|z_P|_p = |\psi_p(P)|_p.$$

We return to our $S$-integral point $P$ which we write as

$$P = p_1 P_1 + \cdots + p_r P_r + T$$

with $T \in \mathrm{Tors}(E)$. We set $Q_i = [m]P_i$ for $i \in \{1, \ldots, r\}$ so that $Q_i \in E_1(\mathbb{Q}_p)$, and hence the $p$-adic elliptic logarithm of $Q_i$ is defined. We then have the inequality

$$\begin{aligned} |p_1 \psi_p(Q_1) + \cdots + p_r \psi_p(Q_r)|_p &= |\psi_p([m]P)|_p \text{ as } [m]P = Q \\ &\leq \frac{1}{\sqrt{|x([m]P)|_p}} \leq \frac{1}{\sqrt{|x(P)|_p}} \\ &\leq \sqrt{c_{11}} e^{-c_{12} K^2/2}. \end{aligned}$$

Hence if we had a very large upper bound on $K$ then we could reduce it using the reduction technique for $p$-adic linear forms in Section VI.4. If we had a lower bound on $|\psi_p([m]P)|_p$ of the standard form for transcendence results then we could deduce an upper bound on $K$ and hence find all the $S$-integral points on the curve.

Such a lower bound has been given in the special case that the elliptic curve has complex multiplication (CM). For those of you are not familiar with CM elliptic curves all you need know is that this really is a very special case. For the case of such CM curves Bertrand [**10**] establishes that there is a constant $c_{13}$ such that

$$-c_{13}(p\log(m^2H))^{16r^2}(\log K)^{8r} < \log|\psi_p([m]P)|_p,$$

where $H$ denotes the maximum of the canonical heights of $P_1, \dots, P_r$. However, the value of the constant is not explicit, but at least it only depends on the coefficients of the curve and the rank. For our application above we would require an explicit value of this constant. At the time of writing, this has only been given for the case $r = 2$ by Rémond and Urfels [**157**].

In the absence of such a result we can, however, use the method above combined with the $p$-adic reduction techniques mentioned earlier to compute all the $S$-integral points on a curve which have height less than some huge constant. For examples of this method see [**179**] and [**76**].

In [**75**] the authors show how one can combine the theoretical estimates of the sizes of $S$-integral points on elliptic curves [**95**] to deduce an upper bound on $K$. Given an upper bound on $K$ we can then use elliptic logarithms to reduce the bound to something very small. Thus they bypass the need for lower bounds for linear forms in $p$-adic elliptic logarithms. Such linear forms are still used in the reduction step, but not in the production of the original bound.

## XIII.6. Other methods and problems

The methods involving elliptic logarithms in this chapter can be extended to elliptic curves defined over algebraic number fields [**185**]. Here an integral point is one whose coordinates live in the ring of integers of a number field. The elliptic logarithms can no longer be forced to be real and hence one obtains a linear form in complex elliptic logarithms. In addition we cannot now use Zagier's method to compute the elliptic logarithms.

The main problem with number fields is computing the Mordell–Weil group. Both the direct method (and descent via 2-isogeny) work but the indirect method is hard to generalize, since it needs an explicit theory of reduction for quartic forms defined over a number field. There has been some work in this direction by Serf and Cremona, for real quadratic fields of class number one [**166**] and [**41**].

Tzanakis [**201**] has extended the method of elliptic logarithms to find all integral points on curves of the form

$$C : Y^2 = aX^4 + bX^3 + cX^2 + dX + e \qquad \text{(XIII.5)}$$

where $C$ has a known rational point. It would be interesting to do the same for other curves of genus one with a known rational point. For instance curves of the form

$$C(X, Y, 1) = 0$$

where $C(X, Y, Z)$ is a ternary cubic form with a known rational point, or

$$Q_1(x_1, x_2, x_3, 1) = Q_2(x_1, x_2, x_3, 1) = 0$$

where the $Q_i$ are quarternary quadratic forms such that $Q_1 \cap Q_2$ is a curve of genus one with a known rational point. We can of course use other methods to determine integral points on curves of the form (XIII.5), for instance reduction to Thue equations, see [**193**], [**210**] and [**212**].

## XIII.7. Exercises

1). Prove that the $AGM$ converges for two positive real numbers $a$ and $b$.

2). Fill in all the messy algebra in showing that $I(a, b) = I(a_1, b_1)$.

3). For each curve in Chapter XII, Exercise 4 for which you know generators of $E(\mathbb{Q})/2E(\mathbb{Q})$ compute all the integral points on the curve.

4). Prove Lemma XIII.5 that if $P$ is an $S$-integral point written in terms of the generators as

$$P = p_1 P_1 + \cdots + p_r P_r + T$$

and $K = \max |p_i|$ then there are positive computable constants $c_{11}$ and $c_{12}$ such that if $P$ is an integral point then for some $v \in S$ we have

$$\frac{1}{|x(P)|_v} \leq c_{11} e^{-c_{12} K^2}.$$

CHAPTER XIV

# Curves of genus greater than one

In this chapter we shall look at an area which has attracted a growing amount of attention in recent years, namely work on curves of genus greater than one. This has come about as the theory of elliptic curves (or curves of genus one) has reached a remarkably advanced state. Work with curves of genus two has been proceeding for over a century but only in the last few years have there been a lot of arithmetical calculations with such curves. There are three basic diophantine questions one can ask about a curve of genus greater than one:

1. Can one compute $C(\mathbb{Q})$, the set of rational points? By a theorem of Faltings this set is known to be finite.
2. Can one compute $C(\mathbb{Z})$, the set of integral points?
3. Can one compute the explicit structure of the Jacobian of $C$? The Jacobian is an algebraic group which is associated with the curve. We shall return to this below.

Clearly if we can give an efficient algorithmic answer to the first question then we already have an efficient algorithm for the second.

For the current state of knowledge on curves of genus two the reader should consult [**26**] and [**156**]. Curves of higher genus have attracted attention from theoreticians but little is known of how to solve the basic diophantine problems on such curves efficiently. We shall start with hyperelliptic curves and then consider various other types of curves. Given the nature of the subject matter we will content ourselves with a short survey and not go into any of the details.

## XIV.1. Curves and their Jacobians

The following section gives a brief introduction to curves and their Jacobians. For more detailed exposition the reader should consult a text on algebraic geometry such as [**97**], the first two chapters of [**172**] or [**136**]. In the next section we will give a more down to earth definition of the Jacobian of a hyperelliptic curve which is suitable for machine computations.

Let $C$ be a projective plane curve of genus greater than one defined over $\mathbb{Q}$ which is the projective normalization of the affine curve given by $C(x, y) = 0$. In other words $C$ is given by some polynomial equation, say $C(x, y) = 0$, in two variables with rational coefficients. A *divisor* on $C$ is a formal finite sum of points of $C$, where the points can be defined over $\overline{\mathbb{Q}}$. We write

$$D = \sum_{P_i \in C} m_i P_i,$$

where $m_i \in \mathbb{Z}$ are called the multiplicities of the points $P_i$, and all but finitely many of the $m_i$ are zero. We define the degree of a divisor $D$ as

$$\deg(D) = \sum m_i.$$

The divisors form a group, $Div(C)$, with respect to the obvious addition law

$$\left(\sum_{P_i \in C} m_i P_i\right) + \left(\sum_{P_i \in C} n_i P_i\right) = \left(\sum_{P_i \in C} (m_i + n_i) P_i\right).$$

An important subgroup is the subgroup of divisors of degree zero which is denoted by $Div^0(C)$.

Every curve gives rise to a field, $\overline{\mathbb{Q}}(C)$, called the function field of $C$. This is the algebraic extension of $\overline{\mathbb{Q}}[x]$ given by

$$\overline{\mathbb{Q}}(C) = \overline{\mathbb{Q}}[x, y]/(C(x, y)).$$

An element $f \in \mathbb{Q}(C)$ is called a 'function on the curve'. To such a function we can associate a divisor called $\mathrm{div}(f)$. This divisor consists of the zeroes and poles of $f$, which lie on the curve with the associated multiplicities. The degree of a divisor of a function is equal to zero as the number of zeroes must be equal to the number of poles when they are counted with the correct multiplicities. A divisor which is the divisor of a function will be called a principal divisor.

The arithmetic of the function field $\overline{\mathbb{Q}}(C)$ shares a remarkable similarity with the arithmetic of an algebraic number field, $K$. The group of divisors of degree zero is analogous to the group of fractional ideals of a number field. Just as in number fields we look at equivalence classes of ideals modulo principal ideals, for a function field we look at the equivalence classes of divisors modulo principal divisors. Two divisors, $D_1$ and $D_2$, will be called equivalent if there is a function, $f \in \overline{\mathbb{Q}}(C)$, such that

$$D_1 = D_2 + \mathrm{div}(f).$$

The group of elements in $Div(C)$ modulo principal divisors is called the Picard group of $C$, $Pic(C)$, whilst the group of divisors of degree zero modulo principal divisors is denoted $Pic^0(C)$.

By the Abel–Jacobi map the group $Pic^0(C)$ is isomorphic to an algebraic variety of dimension $g$. This is just a generalization of the maps from an elliptic curve to the torus, $\mathbb{C}/\Lambda$, given by elliptic integrals. This abelian variety is called the Jacobian of the curve and we shall denote it by $Jac(C)$. In addition the map $Pic^0(C) \to Jac(C)$ commutes with the action of the absolute Galois group, $G = Gal(\overline{\mathbb{Q}}/\mathbb{Q})$. It therefore makes sense to study the set of divisor classes in $Pic^0(C)$ which are fixed by $G$, or equivalently the set of elements in $Jac(C)$ which are fixed by $G$. These two subgroups we shall denote by $Pic^0_{\mathbb{Q}}(C)$ and $J_{\mathbb{Q}}(C)$ respectively. They are the natural objects that one studies when looking at the arithmetic properties of a curve and its Jacobian. Note that whilst a divisor class may be fixed by the action of $G$, such a divisor class need not contain a divisor which is fixed by $G$. A class which is fixed will be called a rational class, whilst a divisor which is fixed will be called a rational divisor.

## XIV.2. Hyperelliptic curves and their Jacobians

To make life easier we shall now consider the case of hyperelliptic curves of genus $g$. Such a curve is given by an equation of the form

$$Y^2 = F(X)$$

where $F$ is a polynomial of degree $n = 2g+1$ or $2g+2$, with no repeated roots. Clearly this is analogous to the short Weierstrass form of an elliptic curve. In what follows we shall let $D_F$ denote the discriminant of the polynomial $F$. Unlike elliptic curves the points on a curve of genus greater than one do not form a group. However, there is a group which is analogous to the Mordell–Weil group of an elliptic curve, namely the Jacobian of the curve.

The Jacobian, $J_{\mathbb{Q}}(C)$, of the curve $C : Y^2 = F(X)$ has various representations; we shall use the following, for the case when the degree of $F$ is equal to $2g + 1$ (for the case when the degree of $F$ is even slight modifications will need to be made). An element of $J_{\mathbb{Q}}(C)$ is represented by two polynomials, $a, b \in \mathbb{Q}[X]$, such that the degree of $a$ is less than $g$, $a$ is monic and $b$ is a polynomial of degree less than the degree of $a$ such that

$$b^2 \equiv F \pmod{a}.$$

A moment's thought will reveal that if $a$ has roots $x_1, \ldots, x_t$ then the points $(x_i, b(x_i))$ all lie on the curve and as a set are invariant under the action of the Galois group. In terms of our previous definition we are representing the elements of $J_{\mathbb{Q}}(C)$ as the degree zero divisor,

$$D = \sum_{i=1}^{\deg a} (x_i, b(x_i)) - (\deg a)\infty.$$

This clearly has degree zero and remains fixed when acted upon by the absolute Galois group. Another thing we notice is that the set of all rational points on $C$ is contained in $J_{\mathbb{Q}}(C)$. If one looks at the above definition in the case when $C$ is an elliptic curve then we see that the elements of the Jacobian are in one to one correspondence with the points on the curve.

We have said that the Jacobian is a group, so we shall need a method to add elements. To add elements there is an algorithm of Cantor [**20**] which we shall now explain. Suppose $(a_1, b_1)$, $(a_2, b_2)$ are two elements of the Jacobian; to compute their sum we perform the following algorithm:

**Adding in the Jacobian: Cantor's algorithm**

DESCRIPTION: Compute $(a_3, b_3) = (a_1, b_1) + (a_2, b_2)$.
INPUT: Polynomials $a_1, a_2, b_1, b_2$.
OUTPUT: Polynomials $a_3, b_3$.

1. Perform two extended gcd computations to compute
   $d = \gcd(a_1, a_2, b_1 + b_2) = h_1 a_1 + h_2 a_2 + h_3(b_1 + b_2)$ .
2. Compute $a_3 = a_1 a_2 / d^2$.
3. Set $b_3 = (h_1 a_1 b_2 + h_2 a_2 b_1 + h_3(b_1 b_2 + F))/d \pmod{a_3}$.
4. While $(\deg a_3 > g)$
5.     $a_3 = (F - b_3^2)/a_3$.
6.     $b_3 = -b_3 \pmod{a_3}$.
7. End while

Such an operation clearly is commutative with the identity given by $(1, 0)$. That it is associative is the hard thing to prove (as it was in the case of elliptic curves). If the degree of $F$ is 3 then the addition law above is identical to the group law on the elliptic curve once we have made the identification of elements of the Jacobian with points on the curve. Because of this we say that an elliptic curve is its own Jacobian.

We can describe the Jacobian of a curve of genus greater than one as a projective variety, and then describe the group law as algebraic maps between points on this variety. This, however, is very cumbersome. For example in genus two when the degree of $F$ is 5 the projective variety lies in $\mathbb{P}^8$, see [**82**], while when the degree is 6 the variety lies in $\mathbb{P}^{15}$ and is described by 72 quadratic forms [**57**]. The definition of the group law in such a situation is also rather complicated, see [**58**].

We can also give the group law for elements in the Jacobian of a curve of genus two in a completely geometric way. First we notice that every element can be expressed as an unordered pair of points on the curve, (which could include the points at infinity). We insist that such a pair is fixed by the action of the absolute Galois group, hence each pair is either a pair of rational points or a pair of quadratic conjugate points. We then 'blow down' pairs of the form $\{(x, y), (x, -y)\}$ to the canonical divisor $\mathcal{O}$. Three pairs of points will then 'add up' to $\mathcal{O}$ if we can find a curve of the form

$$y = a_0 x^3 + a_1 x^2 + a_2 x + a_3$$

which passes through all six component points.

Many of the results in the theory of elliptic curves go over to the theory of Jacobians of curves of genus greater than one. The proofs, however, become far more involved. For example:

THEOREM XIV.1 (Mordell–Weil). *Let $C$ be a curve of positive genus defined over $\mathbb{Q}$, then the group $J_{\mathbb{Q}}(C)$ is a finitely generated abelian group.*

The proof of this result when $C$ is a hyperelliptic curve is rather like the method for elliptic curves. First one proves that the group $J_{\mathbb{Q}}(C)/2J_{\mathbb{Q}}(C)$ is finite. As for elliptic curves this is done using a map from $J_{\mathbb{Q}}(C)$ to the algebra $\mathbb{Q}[X]/(F(X))$. Alas, it is also this part of the method which is not guaranteed to terminate. Given generators of $J_{\mathbb{Q}}(C)/2J_{\mathbb{Q}}(C)$ one can then determine generators of $J_{\mathbb{Q}}(C)$ using an 'infinite descent' argument.

The computation of $J_{\mathbb{Q}}(C)/2J_{\mathbb{Q}}(C)$ for curves of genus two or more, and hence the computation of the rank of $J_{\mathbb{Q}}(C)$, has been carried out for numerous examples in the literature, see for example [**81**], [**59**], [**162**], [**163**] and [**62**]. The infinite descent step has also been carried out for a few examples in genus two, see [**63**].

The outline of the basic method for proving that $J_{\mathbb{Q}}(C)/2J_{\mathbb{Q}}(C)$ is a finite group is as follows, for more details see [**23**] and [**162**]. We shall assume that $F$ is monic of degree $2g+1$ and will let $L$ denote the algebra $\mathbb{Q}[T]/(F(T))$. There is then an injective homomorphism

$$J_{\mathbb{Q}}(C)/2J_{\mathbb{Q}}(C) \rightarrow L^*/L^{*2}$$

given by sending the element $(a,b) \in J_{\mathbb{Q}}(C)$ to the element $a(T)$. We have to have a special fix for elements of the form $(a,0)$ but that will not bother us here, for details of this see [**23**]. The image of this map must then lie in the kernel of the norm map from $L^*/L^{*2}$ to $\mathbb{Q}^*/\mathbb{Q}^{*2}$. Indeed, using the same sort of argument as we did for the direct method for elliptic curves, we see that the image lies in a direct sum of groups of the form $K_i(S,2)$, where $K_i$ is a number field and $S$ is the set of primes dividing $2D_F$. Hence the group $J_{\mathbb{Q}}(C)/2J_{\mathbb{Q}}(C)$ is finite as its image under the above injective homomorphism is finite. The main problem, as in the case of elliptic curves, is effectively determining the image of the above homomorphism.

## XIV.3. Rational points on curves of genus greater than one

Mordell had originally conjectured that on a curve of genus greater than one there are only finitely many rational points. In 1983 Faltings, [**55**], proved this result, in an ineffective way. Faltings showed that if $C$ was a curve of genus greater than one and $K$ was a number field of finite degree over $\mathbb{Q}$ then the set of $K$-rational points on $C$, $C(K)$, was a finite set. There have been many attempts to make Faltings' method effective but none have succeeded in complete generality.

The first check that we need to carry out on any given curve, $C$, is that it has solutions in every local field $\mathbb{Q}_p$. This allows us to eliminate a curve immediately in a very quick manner. The methods for doing this are analogous to those we have looked at before. Namely finding a solution modulo a high enough power of $p$ and then lifting it using Hensel's lemma to a solution in $\mathbb{Q}_p$.

If we are lucky we can compute the complete set of rational solutions on some curve of genus greater than one using a result of Chabauty [**28**].

Chabauty showed that if the rank of the Mordell–Weil group of the Jacobian was less than the genus of the curve then the number of rational points was finite. Coleman [**35**] showed using Chabauty's method that if $C$ is a curve of genus $g$ whose Jacobian has rank less than $g$ and $p > 2g$ is a prime such that the curve has 'good reduction' modulo $p$ then

$$|C(\mathbb{Q})| \leq |C(\mathbb{F}_p)| + 2g - 2.$$

For the case of hyperelliptic curves the condition that $C$ must have good reduction modulo $p$ means that $p$ does not divide $2D_F$.

Further refinements to this method have been developed by Flynn [**60**] and [**61**] for curves of genus two. In genus two the formal group is given by two power series in two variables, just as for elliptic curves the formal group was given by one power series in one variable. For curves of genus two with Jacobians of rank one, Flynn produces a condition which the power series must satisfy for a rational point to exist. The number of solutions to the $p$-adic power series can then be bounded using Strassmann's theorem, Theorem II.5. If we are lucky then this bound is sharp enough to allow us to provably write down all the rational points. If we are not lucky we still have a chance by using different prime. Using this method we can prove, see [**26**], examples such as

LEMMA XIV.2. *The only rational points on the curve*

$$Y^2 = 2X(X^2 - 2X - 2)(1 - X^2)$$

*are* $(0,0), (\pm 1, 0), (-1/2, \pm 3/4)$ *and the point at infinity.*

Some other methods to find all the rational points on a curve of genus greater than one involve linking the rational points on the curve $C$ to some properties of another curve $D$. For example suppose we wish to find the rational points on a curve $C$. If we can find a non-constant morphism, $C \to D$, to another curve, $D$, then determining the set of rational points on $D$, if finite, will allow us to find the set rational points on $C$.

As an example suppose $C$ is a hyperelliptic curve of genus two of the form

$$C : Y^2 = aX^6 + bX^4 + cX^2 + d,$$

then we can let $D$ denote the elliptic curve $Y^2 = aX^3 + bX^2 + cX + d$. We would clearly have the map

$$\begin{array}{ccc} C & \longrightarrow & D \\ (X,Y) & \longmapsto & (X^2, Y) \end{array}.$$

Now if $D$ had rank zero we could determine all the elements of $D(\mathbb{Q})$, which would be finite. Determining $C(\mathbb{Q})$ is then easy as all we need do is determine which of the elements of $D(\mathbb{Q})$ have an $x$-coordinate which is a rational square.

This trick is very special but it has been extended to a more general method by Dem'janenko and Manin [**46**], [**125**]. Their method is applicable when the curve under consideration, like the one above, has maps into some elliptic curve $E$. If the rank of the group of such morphisms is greater than the rank of the elliptic curve $E$ then the number of rational points on $C$ can be proved to be finite. Moreover such a proof is effective.

Yet another method has been given by Coombes and Grant [**36**]. In this method curves $D$ are constructed which are unramified covers of $C$, $\phi : D \rightarrow C$. In such a situation there is a number field $K$ such that $C(\mathbb{Q}) \subset \phi(D(K))$. Hence determining the $K$-rational points on all such $D$ may allow one to determine the $\mathbb{Q}$-rational points on $C$. If the genus of $C$ is greater than one then the genus of $D$ will be strictly larger than the genus of $C$. This appears to have made the problem harder. However, $D$ may possess maps down to some elliptic curve $E$, for which we can apply the method of Dem'janenko and Manin. Or maybe we can apply the method of Chabauty to determine the finite set of $K$-rational points on $D$.

Using this method Coombes and Grant show that if $p$ is a prime which is congruent to 7 modulo 16 then the only rational points on

$$y^2 = x^5 + px$$

are given by $(0, 0)$ and the point at infinity.

We shall not go into detail into these methods for finding rational points as they are slightly more advanced than many of the others in this book, and most of them require a knowledge of algebraic geometry which I have not assumed. Luckily the problem of finding all integral points on such curves is easier to understand as it is based on methods we have already met. However, it should be borne in mind that the majority of the methods for rational points, when they work, are much simpler to apply than the methods for finding all the integral points. It hardly needs to be pointed out that if one has a quick method to find all the rational points on a curve then there seems no point in applying a complicated method to find all the integral points.

## XIV.4. Integral points on hyperelliptic and superelliptic curves

Luckily we already have a method to find integral points on a hyperelliptic curve as the method of Section IX.4 can be applied word for word. However, as pointed out in that section the method is not very efficient as the method requires solving unit equations in large degree number fields.

In the special case when our curve is of the form

$$Y^2 = X^n + c,$$

where $c$ is not a square, there is a much better method. We rewrite the equation as

$$X^n = Y^2 - c = (Y - \sqrt{c})(Y + \sqrt{c}).$$

Then if we set $K = \sqrt{c}$, then for a finite set of 'bad' primes, $S$, we have

$$Y - \sqrt{c} = \alpha(x_1 + \sqrt{c}x_2)^n$$

where $\alpha \in K(S,n)$. Equating coefficients of $\sqrt{c}$ in both sides gives us the equations

$$\begin{aligned} F_1(x_1,x_2) &= Y, \\ F_2(x_1,x_2) &= -1, \end{aligned}$$

where $F_1(x_1,x_2)$ and $F_2(x_1,x_2)$ are binary forms of degree $n$. The last of these is a Thue equation and hence can be solved by the method of Chapter VII. Having solved for $(x_1,x_2)$ we can then use the first equation to determine the value of $Y$. For an example of this method see [**133**], [**213**]. For other methods of solving hyperelliptic equations see [**209**], [**13**].

Whilst we are looking at hyperelliptic equations we may as well study their near relation, the case of superelliptic equations. A superelliptic equation is one of the form

$$Y^n = F(X)$$

where $F(X)$ is a polynomial with at least two distinct roots of multiplicity one and $n$ is an integer greater than or equal to three. We can easily reduce this equation to the solution of a set of Thue equations defined over a number field in the following manner.

Let $\alpha, \beta$ denote two distinct roots of $F(X) = 0$ of multiplicity one and set $K_1 = \mathbb{Q}(\alpha)$ and $K_2 = \mathbb{Q}(\beta)$. A by now standard analysis gives us that for a solution $(X,Y)$ to our superelliptic equation we must have

$$\begin{aligned} X - \alpha &= \sigma_1 \tau_1^n, \\ X - \beta &= \sigma_2 \tau_2^n, \end{aligned}$$

where $\tau_i \in \mathcal{O}_{K_i}$ and $\sigma_i \in K_i(S,n)$ for some finite sets of primes $S_i$. Substituting one of these into the other then gives us

$$\sigma_1 \tau_1^n - \sigma_2 \tau_2^n = \beta - \alpha$$

which is a Thue equation for $(\tau_1,\tau_2)$ defined over the number field $K = \mathbb{Q}(\alpha,\beta)$. As Thue equations are triangularly connected these Thue equations can be solved by the method in Chapter X. This is hardly an efficient and easy method to implement.

Bilu and Hanrot [**13**] determine a more efficient way of computing all the integral points on a superelliptic curve. Their method makes use of the same two equations above involving $\alpha, \beta, \sigma_1$ and $\sigma_2$, but they do not reduce the problem to a Thue equation over $K$. Instead with careful analysis of various number fields they directly produce a linear form in logarithms which the standard results can be applied to.

It is also worth noting that there has been a lot of theoretical work on bounding the number of solutions to elliptic, hyperelliptic and superelliptic equations, see for instance [**54**], [**84**] and [**205**].

## XIV.5. Fermat curves

One famous class of curves is the Fermat curves,

$$x^n + y^n = 1.$$

In his famous marginal note Fermat stated that he had a proof that the only non-trivial rational solutions on such curves occurred when $n$ was equal to two. That there were infinitely many non-trivial solutions when $n$ was equal to two had been known since antiquity. However, no record of Fermat's proof, if one actually existed, has ever been found.

When studying Fermat's equation it is clear that if $n = pq$ has a non-trivial solution, $(x, y)$, then $(x^q, y^q)$ will be a non-trivial solution of $x^p + y^p = 1$. As it had been known since the time of Fermat that Fermat's last theorem was true when $n = 4$, Theorem I.1, all that was left was to prove the theorem when $n$ was an odd prime number.

Multiplying up by a common denominator we are led to look at the integral solutions to the homogeneous problem,

$$x^p + y^p = z^p, \tag{XIV.1}$$

when $p$ is a prime greater than 4, the case $p = 3$ having been proved by Euler. Much work had been done on this equation over the centuries but it was only in recent years that a proof was discovered by Wiles [**217**].

The main idea is far removed from the subject of this book, but a book on diophantine equations would seem incomplete without even a brief outline. Frey [**64**] had noticed that if a non-trivial coprime solution $(x, y, z) = (a, b, c)$ existed to (XIV.1) then one could form the elliptic curve

$$E : Y^2 = X(X - a^p)(X + b^p),$$

a curve which is now called a Frey curve. We can assume without loss of generality that $b$ is even and $a \equiv -1 \pmod 4$. If we look at the discriminant of the cubic polynomial on the right hand side we see that it is equal to

$$D = (0 - a^p)^2(0 - b^p)^2(a^p + b^p)^2 = (abc)^{2p}.$$

This led Frey to believe that such a curve would have to be very weird indeed. For example the curve is semistable and has conductor

$$N = \prod_{q|abc} q,$$

its minimal discriminant is $2^{-8}D$, whilst its $j$-invariant is given by

$$2^8 \frac{a^{2p} + b^{2p} + a^p b^p}{D}.$$

Notice that for every odd prime $q$ dividing $N$ we then have that $q$ divides $j$ to a $p$th power exactly.

Using ideas of Serre, Ribet [**158**] showed that if such a Frey curve existed then it really would be a weird curve as it would provide a counterexample to the conjecture of Shimura–Taniyama–Weil. This was a deep conjecture which linked the theory of elliptic curves to the theory of modular functions of one variable.

Wiles then set about trying to prove the Shimura–Taniyama–Weil conjecture. In 1995 Wiles [**217**] finally published a proof of enough of the conjecture to show that Frey's elliptic curve could not exist. If Frey's elliptic curve could not exist, as it violated the Shimura–Taniyama–Weil conjecture, then this meant that the original non-trivial solution to the Fermat equation could also not exist. Hence Fermat's last theorem had been proved.

A number like $N$ above also occurs in a conjecture which is related to another proposed proof of Fermat's last theorem;

CONJECTURE XIV.3 (ABC-conjecture). *Given $\epsilon > 0$ there is a constant $C$ such that for all non-zero relatively prime integers, $a, b, c$, such that $a + b = c$ we have*

$$\max(|a|, |b|, |c|) \leq C \left( \prod_{q|abc} q \right)^{1+\epsilon}.$$

## XIV.6. Catalan's equation

Related to the Fermat curves in the last section is Catalan's equation,

$$x^m - y^n = 1.$$

In 1844 E. Catalan [**27**] conjectured that this equation had only the trivial solution $(x, y, m, n) = (3, 2, 2, 3)$. Lebesgue [**114**] showed that there were no solutions when $n = 2$ and Nagell [**139**] showed there was only the trivial solution when either $m = 3$ or $n = 3$. In addition Chao Ko [**106**] showed that when $m = 2$ there was only the trivial solution. So clearly, in an attempt to prove Catalan's conjecture, we may assume that $\min(m, n) \geq 5$.

It is often convenient when looking at Catalan's equation to perform the obvious reduction to the case where $n$ and $m$ are prime, just as we did for the Fermat curves above. We also usually assume that if $p$ and $q$ are the two prime exponents then $p > q$. This means we need to show that there are no solutions to the equation

$$x^p - y^q = \pm 1 = \epsilon$$

if $p > q \geq 5$. When $p$ and $q$ are fixed then we have nothing to do but show that the equation has no rational solutions $(x, y)$. But when $p$ and $q$ are fixed this is nothing but a superelliptic equation. So at least we know that there are only finitely many possible values of $(x, y)$ in this case. Cassels proved [**21**] that we must have $p|y$ and $q|x$. Alas proving there are no values for $x$ and $y$ would involve a lot of computing for even smallish values of $p$ and $q$ so our attention must clearly shift to eliminating various values for the exponents.

Life would certainly appear easier if we knew there were only finitely many values of $p$ and $q$ which we need to check. If this were true then we would know that Catalan's equation had only finitely many solutions. Luckily such a result is true, as was first proved by Tijdeman [**197**] in 1976. Various improvements have been made to Tijdeman's method, for instance see [**80**]; we shall only give a proof outline.

By looking at the factorizations of $y^q+\epsilon$ and $x^p-\epsilon$, and using the result of Cassels mentioned previously, we can deduce that there are integers $s$ and $r$ with $q|s$ and $p|r$ such that

$$\begin{aligned} y+\epsilon &= s^p/q, \\ x-\epsilon &= r^q/p. \end{aligned}$$

Various inequalities can then be established which demonstrate that $r$ and $s$ are very nearly equal in size. Combining these inequalities with a lower bound on the linear form in logarithms

$$\Lambda_1 = |pq\log(r/s) + q\log q - p\log p| \leq 4p^q/r^q$$

leads one to deduce that, for some explicit positive constant $c_1$,

$$q \leq c_1(\log p)^3.$$

We then turn our attention to the linear form in logarithms

$$\Lambda_2 = \left| p\log\left(\frac{r^q/p+\epsilon}{s^q}\right) + q\log q \right|.$$

The various inequalities allow us to deduce $\Lambda_2 \leq 4q^q/s^p$ and so, on applying a lower bound on the above linear form, we can deduce that $p \leq c_2$, for some explicit positive constant $c_2$. Using the best available lower bounds on linear forms in two and three logarithms, which are often better than the general case, we can obtain

THEOREM XIV.4 ([**80**]). *If $p$ and $q$ are prime numbers with $p > q \geq 5$ and $x$ and $y$ are integer solutions to*

$$x^p - y^q = \epsilon$$

*then $p \leq 3.42 \cdot 10^{28}$ and $q \leq 6.0 \cdot 10^{19}$.*

However, $p$ and $q$ are still far too large to allow a brute force attack on the problem. In previous examples we have used the LLL–algorithm to reduce such astronomical upper bounds. However, in all the previous examples we had actual numbers in the linear forms in logarithms. In the case of Catalan's equation the linear forms, $\Lambda_1$ and $\Lambda_2$, consist of logarithms of unknown quantities. It is for this reason that we cannot reduce the upper bounds.

Some progress can be made in eliminating various pairs of primes in this range from further consideration. There are various results along these lines, see [**102**] and [**103**]. For example if we let $h_q$ denote the class number of $\mathbb{Q}(\zeta_q)$ and $h(-q)$ denote the class number of $\mathbb{Q}(\sqrt{-q})$ then we can show:

THEOREM XIV.5. *If $p$ and $q$ are prime numbers with $p > q$ then there are no solutions to $x^p - y^q = \epsilon$ if $q^{p-1} \equiv 1 \pmod{p^2}$ and one of the following conditions holds:*

1. *$p$ does not divide $h_q$.*
2. *$q \equiv 3 \pmod 4$ and $p$ does not divide $h(-q)$.*

Results like this, and those in [**165**], have been used to eliminate a large number of possible values for $(p, q)$. See [**80**], [**131**], [**134**] and [**135**] where it is shown that if there is a non-trivial solution to Catalan's equation then $p > 10^6$ and $q > 10^5$.

## XIV.7. Exercises

1). Show that the ABC-conjecture implies that Fermat's last theorem is true for all sufficiently large exponents.

2). Prove the following polynomial version of the ABC-conjecture. If $a, b, c$ are relatively prime polynomials such that $a + b = c$ then

$$\max(\deg a, \deg b, \deg c) \leq (\text{ the number of distinct roots of } abc) - 1.$$

3). Determine explicit values of the constants $c_1$ and $c_2$ in the proof outline of Theorem XIV.4.

4). Determine all the rational points on the curve

$$Y^2 = X^{63} + 124.$$

APPENDIX A

# Linear forms in logarithms

We now present the three transcendence results we need in other parts of this book. Firstly there is one for standard complex logarithms of algebraic numbers, then there is one which amounts to the same for $p$-adic logarithms of algebraic numbers and finally we give one for elliptic logarithms. In what follows $\log(\alpha)$ will always denote the principal value of the complex logarithm of a complex number.

## A.1. Linear forms in complex logarithms

Let $\alpha_1, \ldots, \alpha_n$ (with $n \geq 2$) denote algebraic numbers not equal to 0 or 1. Let $K = \mathbb{Q}(\alpha_1, \ldots, \alpha_n)$ and set $d = [K : \mathbb{Q}]$. We define a modified height by the formula

$$h_m(\alpha) = \max\left\{h(\alpha), \frac{|\log \alpha|}{d}, 1/d\right\}.$$

THEOREM A.1 (Baker-Wüstholz [**6**]). *Let $b_1, \ldots, b_n$ be integers such that*

$$\Lambda = b_1 \log \alpha_1 + \cdots + b_n \alpha_n$$

*is non-zero. Then if $B = \max\{|b_1|, \ldots, |b_n|\} \geq 3$ we have the inequality*

$$-c_1 h_m(\alpha_1) \cdots h_m(\alpha_n) \log B < \log |\Lambda|$$

*with*

$$c_1 = 18(n+1)!n^{n+1}(32d)^{n+2} \log(2nd).$$

There are better results for linear forms in exactly two or three logarithms but the above is the best general result at the time of writing. In this book we do not make use of the special results for two and three logarithms as this may confuse. You should consider the use of linear form results in two and three logarithms as nothing but an algorithmic optimization. It is better to understand the basic method first before worrying about such optimizations. In fact in practice we will still need to use LLL to reduce the upper bounds so very little is gained from the use of such results.

## A.2. Linear forms in $p$-adic logarithms

Let $\alpha_1, \ldots, \alpha_n$ (with $n \geq 2$) denote non-zero algebraic numbers. Let $K = \mathbb{Q}(\alpha_1, \ldots, \alpha_n)$ and set $d = [K : \mathbb{Q}]$. As usual we let $p$ denote a rational prime,

$\mathfrak{p}$ a prime ideal lying above $p$ in $K$ with residue degree $f_{\mathfrak{p}}$. We now define a modified height by the formula

$$h_m(\alpha) = \max\left\{h(\alpha), \frac{|\log \alpha|}{2\pi D}, \frac{f_{\mathfrak{p}} \log p}{d}\right\},$$

where $D = d$ if $p > 2$ and $\sqrt{-1} \in K$ or $p = 2$ and $K$ contains a cube root of unity, in all other cases $D = 2d$.

THEOREM A.2 (Yu, see [**218**] and [**203**, Appendix] ). *Suppose* $\text{ord}_{\mathfrak{p}}(\alpha_j) = 0$ *for all* $j = 1, \ldots, n$ *and let* $b_1, \ldots, b_n$ *be integers such that* $\alpha_1^{b_1} \cdots \alpha_n^{b_n} \neq 1$, *then if we set* $B = \max\{|b_1|, \ldots, |b_n|\}$ *we have the inequality*

$$\text{ord}_{\mathfrak{p}}\left(\alpha_1^{b_1} \cdots \alpha_n^{b_n} - 1\right) < c_2 c_3 c_4 (\log B + c_5)$$

*where*

$$\begin{aligned} c_2 &= \begin{cases} 35009 \cdot (45/2)^n & \textit{if } p \equiv 1 \pmod 4, \\ 30760 \cdot 25^n & \textit{if } p \equiv 3 \pmod 4, \\ 197142 \cdot 36^n & \textit{if } p = 2. \end{cases} \\ c_3 &= (n+1)^{2n+4} p^{D f_{\mathfrak{p}}/d} (f_{\mathfrak{p}} \log p)^{-(n+1)} D^{n+2} h_m(\alpha_1) \cdots h_m(\alpha_n), \\ c_4 &= \begin{cases} \log(2^{11}(n+1)^2 D^2 H) & \textit{if } p > 2, \\ \log(3 \cdot 2^{10}(n+1)^2 D^2 H) & \textit{if } p = 2. \end{cases} \\ c_5 &= 2 \log D, \\ H &= \max\{h_m(\alpha_1), \ldots, h_m(\alpha_n)\}. \end{aligned}$$

This may not seem to be a result about linear forms in $p$-adic logarithms but it is easy to convert it into such a result using Lemma II.9, the details of which we leave as an easy exercise.

## A.3. Linear forms in elliptic logarithms

We now present David's result [**44**] on linear forms in elliptic logarithms. We slightly simplify the result to something which is more manageable for our purposes. Let $E$ denote the elliptic curve

$$E : Y^2 = 4X^2 - g_2 X - g_3,$$

with invariant $j$ and periods $\omega_1, \omega_2$ such that $\tau = \omega_1/\omega_2$ has positive imaginary part. Define the height of the elliptic curve by

$$h_E = \max(1, h(1, g_2, g_3), h(j)).$$

Let $\{P_1, \ldots, P_n\}$ denote points on $E(K)$ for some number field $K$ of degree $D$ (within which we assume lie $g_2$ and $g_3$). Let $\psi$ denote the elliptic logarithm map

$$\psi(P) = \int_{\infty}^{x + b_2/12} \frac{dt}{\sqrt{4t^3 - g_2 t - g_3}} \pmod \Lambda,$$

where $\Lambda = \omega_1\mathbb{Z} + \omega_2\mathbb{Z}$. Set

$$c_7 = \frac{3\pi}{|\omega_1|^2\Im(\tau)}.$$

Define the modified height of the points $P_i$ by

$$h_m(P_i) = \max\left\{\hat{h}(P_i), h_E, c_7|\psi(P_i)|^2/D\right\}.$$

Set

$$\begin{aligned} c_8 &= \max\{eh_E, h_m(P_1)/D, \ldots, h_m(P_n)/D\}, \\ c_9 &= \min_{i=1\ldots n}\left\{\frac{e\sqrt{Dh_m(P_i)}}{\sqrt{c_7}|\psi(P_i)|}\right\}. \end{aligned}$$

We are now ready to state the theorem:

THEOREM A.3 (David [**44**]). *Set*

$$L(\vec{x}) = \sum_{i=1}^{n} x_i\psi(P_i)$$

*with* $\vec{x} \in \mathbb{Z}^n$. *Set* $A = \max|x_i|$ *then if* $A \geq \exp(c_8)$ *and* $L(\vec{x}) \neq 0$ *we have*

$$\log|L(\vec{x})| > -c_{10}(\log A + \log(Dc_9))(\log\log A + h + \log(Dc_9))^{n+1}$$

*where*

$$c_{10} = 2 \cdot 10^{8+7n}(2/e)^{2n^2}(n+1)^{4n^2+10n}D^{2n+2}(\log c_9)^{-2n-1}\prod_{i=1}^{n} h_m(P_i).$$

APPENDIX B

# Two useful lemmata

In this appendix we give two results which are needed throughout the book. Firstly we give the lemma of Pethő and de Weger, an elementary result which was first given in [**151**]. However, it is used over and over again so it is worth stating explicitly.

LEMMA B.1 (Pethő and de Weger). *Let $a, b \geq 0, h \geq 1$ and $x \in \mathbb{R}$ be the largest solution of $x = a + b(\log x)^h$. If $b > (e^2/h)^h$ then*

$$x < 2^h \left(a^{1/h} + b^{1/h} \log(h^h b)\right)^h$$

*and if $b \leq (e^2/h)^h$ then*

$$x \leq 2^h \left(a^{1/h} + 2e^2\right)^h.$$

PROOF. Put $c = hb^{1/h}$ and define $y$ by $(1+y)c\log c = x^{1/h}$. Notice from the obvious inequality $(z_1 + z_2)^{1/h} \leq z_1^{1/h} + z_2^{1/h}$ that we have

$$\begin{aligned}
(1+y)c\log c &= x^{1/h} = \left(a + b(\log x)^h\right)^{1/h}, \\
&\leq a^{1/h} + b^{1/h}\log x, \\
&= a^{1/h} + \frac{c}{h}\log x, \\
&= a^{1/h} + c\log x^{1/h}, \\
&= a^{1/h} + c\log\left((1+y)c\log c\right), \\
&= a^{1/h} + c\log(1+y) + c\log c + c\log\log c, \\
&< a^{1/h} + cy + c\log c + c\log\log c.
\end{aligned}$$

Hence we can deduce that

$$yc(\log c - 1) < a^{1/h} + c\log\log c.$$

Firstly assume that $b > (e^2/h)^h$ then $c > e^2$ and so

$$\begin{aligned}
x^{1/h} &= c\log c + yc\log c, \\
&< c\log c + \frac{\log c}{\log c - 1}\left(a^{1/h} + c\log\log c\right), \\
&< 2\left(a^{1/h} + c\log c\right).
\end{aligned}$$

which we note will also hold if $c = e^2$.

Now when $b \leq (e^2/h)^h$ we then have

$$x \leq a + (e^2/h)^h (\log x)^h$$

and so we can choose $c = e^2$ and then the above inequality gives

$$x^{1/h} < 2\left(a^{1/h} + 2e^2\right),$$

as required. □

The final result we shall require is the following simple lemma which is used over and over again to produce linear forms in logarithms which are small.

LEMMA B.2. *Let $\Delta \in \mathbb{C}$ with $|\Delta - 1| \leq a$. Then*

$$|\log \Delta| \leq \frac{-\log(1-a)}{a} |\Delta - 1|.$$

PROOF. We have

$$\begin{aligned} |\log \Delta| &= \left| \sum_{i=1}^{\infty} \frac{(-1)^{i-1}(\Delta - 1)^i}{i} \right| \\ &\leq |\Delta - 1| \sum_{i=1}^{\infty} \frac{a^{i-1}}{i} \\ &= \frac{-\log(1-a)}{a} |\Delta - 1|. \end{aligned}$$

In particular if $|\Delta| \leq 0.5$ then $|\log \Delta| \leq 2|\Delta - 1|$. □

# References


[1] L.M. Adleman and M.-D. Huang, editors. *ANTS-1: Algorithmic Number Theory.* Springer–Verlag, LNCS 877, 1994.

[2] M. Agrawal, J. Coates, D. Hunt, and A.J. van der Poorten. Elliptic curves of conductor 11. *Math. Comp.*, **35**, 991–1002, 1980.

[3] E. Bach and J. Shallit. *Algorithmic Number Theory. Volume 1: Efficient Algorithms.* MIT Press, 1996.

[4] A. Baker. Contributions to the theory of diophantine equations I and II. *Phil. Trans. Roy. Soc. London Ser A.*, **263**, 173–208, 1968.

[5] A. Baker and H. Davenport. The equations $3x^2 - 2 = y^2$ and $8x^2 - 7 = z^2$. *Quart. J. Math.*, **20**, 129–137, 1969.

[6] A. Baker and G. Wüstholz. Logarithmic forms and group varieties. *J. Reine Angew. Math.*, **442**, 19–62, 1993.

[7] C. Batut, D. Bernardi, H. Cohen, and M. Olivier. GP/PARI version 1.39.03. *Université Bordeaux I*, 1994.

[8] M.A. Bennett. Solving norm form equations via lattice basis reduction. *Rocky Mountain Journal of Maths*, **26**, 815–837, 1996.

[9] M.A. Bennett and B.M.M. de Weger. On the diophantine equation $|ax^n - by^n| = 1$. *Math. Comp.*, **67**, 413–438, 1998.

[10] D. Bertrand. Approximations diophantiennes $p$–adiques sur les courbes elliptiques admettant et multiplication complexe. *Comp. Math.*, **37**, 21–50, 1978.

[11] Y. Bilu and G. Hanrot. Solving Thue equations of high degree. *J. Number Th.*, **60**, 373–392, 1996.

[12] Y. Bilu and G. Hanrot. Thue equations with composite fields. To appear: *Acta. Arith.*

[13] Y. Bilu and G. Hanrot. Solving superelliptic diophantine equations by Baker's method. Preprint.

[14] B.J. Birch and J.R. Merriman. Finiteness theorems for binary forms with given discriminant. *Proc. L.M.S.*, **24**, 385–394, 1972.

[15] B.J. Birch and H.P.F. Swinnerton-Dyer. Notes on elliptic curves. I. *J. Reine Angew. Math.*, **212**, 7–25, 1963.

[16] E. Bombieri and W.M. Schmidt. On Thue's equation. *Invent. Math.*, **88**, 69–81, 1987.

[17] W. Bosma, J.J. Cannon, and C. Playoust. The Magma algebra system I: The user language. *J. Symbolic Computation*, **24**, 235–265, 1997.

[18] A. Bremner. On the equation $y^2 = x(x^2 + p)$. In R.A. Mollin, editor, *Number Theory and Applications*, pages 3–23. Kluwer, Dordrecht, 1989.

[19] A. Bremner and J.W.S. Cassels. On the equation $y^2 = x(x^2 + p)$. *Math. Comp.*, **42**, 257–264, 1984.

[20] D.G. Cantor. Computing in the Jacobian of a hyper–elliptic curve. *Math. Comp.*, **48**, 95–101, 1987.

[21] J.W.S. Cassels. On the equation $a^x - b^y = 1$, II. *Proc. Camb. Phil. Soc.*, **56**, 97–103, 1960.

[22] J.W.S. Cassels. Diophantine equations with special reference to elliptic curves. *J. of LMS*, **41**, 193–291, 1966.

[23] J.W.S. Cassels. The Mordell–Weil group of curves of genus 2. In *Arithmetic and Geometry Papers Dedicated to I.R.Shafarevich on the Occasion of his Sixtieth Birthday, Vol. 1*, pages 29–60. Birkhäuser, 1983.

[24] J.W.S. Cassels. *Local Fields.* LMS Student Texts, Cambridge University Press, 1986.

[25] J.W.S. Cassels. *Lectures on Elliptic Curves.* LMS Student Texts, Cambridge University Press, 1991.

[26] J.W.S. Cassels and E.V. Flynn. *Prolegomena to a Middlebrow Arithmetic of Curves of Genus 2.* Cambridge University Press, 1996.

[27] E. Catalan. Note extraite d'une lettre adressée à l'éditeur. *J. reine. angew. Math.*, **27**, 192, 1844.

[28] C. Chabauty. Sur les points rationnels des courbes algébriques de genre supérieur à l'unité. *Comptes Rendus Hebdomadaires des Séances de l'Acad. des Sci. Paris*, **212**, 882–885, 1941.

[29] J.H. Chen and P.M. Voutier. Complete solution of the diophantine equation $x^2+1 = dy^4$ and a related family of quartic Thue equations. *J. Number Theory*, **62**, 71–99, 1997.

[30] J. Coates. An effective $p$–adic analogue of a theorem of Thue. *Acta Arith.*, **15**, 279–305, 1969.

[31] F.B. Coghlan and N.M. Stephens. The diophantine equation $x^3 - y^2 = k$. In A.O.L. Atkin and B.J. Birch, editors, *Computers in Number Theory*, pages 199–205. Academic Press, 1971.

[32] H. Cohen. *A Course In Computational Algebraic Number Theory.* Springer–Verlag, GTM 138, 1993.

[33] H. Cohen, editor. *ANTS-2: Algorithmic Number Theory.* Springer–Verlag, LNCS 1122, 1996.

[34] H. Cohen, F. Diaz Y Diaz, and M. Olivier. Subexponential algorithms for class group and unit computations. *J. Symbolic Computation*, **24**, 433–441, 1997.

[35] R.F. Coleman. Effective Chabauty. *Duke Math. J.*, **52**, 765–780, 1985.

[36] K.R. Coombes and D. Grant. On heterogeneous spaces. *J. London Math. Soc.*, **40**, 385–397, 1989.

[37] D.A. Cox. The arithmetic–geometric mean of Gauss. *L'Enseignement Mathématique*, **30**, 275–330, 1984.

[38] D.A. Cox. Gauss and the arithmetic–geometric mean. *Notices AMS*, pages 147–151, 1985.

[39] J.E. Cremona. Classical invariants and 2–descent on elliptic curves. Preprint.

[40] J.E. Cremona. *Algorithms for Modular Elliptic Curves.* Cambridge University Press, 1992.

[41] J.E. Cremona and P. Serf. Computing the rank of elliptic curves over real quadratic fields of class number 1. To appear: *Math. Comp.*

[42] M. Daberkow, C. Fieker, J. Klüners, M. Pohst, K. Roegner, M. Schörnig, and K. Wildanger. KANT V4. *J. Symbolic Computation*, **24**, 267–283, 1997.

[43] H. Davenport. *The Higher Arithmetic.* Cambridge University Press, 1992.

[44] S. David. Minorations de formes linéaires de logarithmes elliptiques. *Mém. Soc. Math. France*, **62**, 1995.

[45] M. Davis, J. Matijasevič, and J. Robinson. Hilbert's tenth problem. diophantine equations: positive aspects of a negative solution. In F. Browder, editor, *Mathematical Developments Arising from Hilbert Problems.* AMS, Providence, 1976.

[46] V. Dem'janenko. Rational points on a class of algebraic curves. *Amer. Math. Soc. Transl.*, **66**, 246–272, 1968.

[47] F. Diamond. On deformation rings and Hecke rings. *Annals of Math*, **144**, 137–166, 1996.

[48] E.B. Elliott. *An Introduction to the Algebra of Quantics*. Oxford University Press, 1895.

[49] W.J. Ellison. Recipes for solving diophantine problems by Baker's method. *Séminaire de théorie des nombres, Université de Bordeaux I (1970–71)*.

[50] V. Ennola. Cubic number fields with exceptional units. In [**152**], pages 103–128.

[51] M. Euchner and C.P. Schnorr. Lattice basis reduction: improved practical algorithms and solving subset sum problems. In *Proc FCT 1991*, LNCS 529, pages 68–85. Springer–Verlag, 1991.

[52] J.H. Evertse and K. Győry. Effective finiteness results for binary forms with given discriminant. *Comp. Math.*, **79**, 169–204, 1991.

[53] J.H. Evertse. On equations in $S$–units and the Thue–Mahler equation. *Invent. Math.*, **75**, 561–584, 1984.

[54] J.H. Evertse and J.H. Silverman. Uniform bounds for the number of solutions to $y^n = f(x)$. *Proc. Camb. Phil. Soc.*, **100**, 237–248, 1986.

[55] G. Faltings. Endlichkeitssästze fűr abelsche Varietäten über Zahlenkőrpen. *Inv. Math.*, **73**, 349–366, 1983.

[56] U. Fincke and M. Pohst. Improved methods for calculating vectors of short length in a lattice, including a complexity analysis. *Math. Comp.*, **44**, 463–471, 1985.

[57] E.V. Flynn. The Jacobian and formal group of a curve of genus 2 over an arbitrary ground field. *Proc. Camb. Phil. Soc.*, **107**, 425–441, 1990.

[58] E.V. Flynn. The group law on the Jacobian of a curve of genus 2. *J. reine. angew. Math.*, **439**, 45–69, 1993.

[59] E.V. Flynn. Descent via isogeny in dimension 2. *Acta. Arith.*, **66**, 23–43, 1994.

[60] E.V. Flynn. On a theorem of Coleman. *Manuscripta Math.*, **88**, 447–456, 1995.

[61] E.V. Flynn. A flexible method for applying Chabauty's theorem. *Comp. Math.*, **105**, 79–94, 1997.

[62] E.V. Flynn, B. Poonen, and E.F. Schaefer. Cycles of quadratic polynomials and rational points on a genus 2 curve. To appear: *Duke Math. Journal.*

[63] E.V. Flynn and N.P. Smart. Canonical heights on the Jacobians of curves of genus 2 and the infinite descent. *Acta. Arith.*, **79**, 333–352, 1997.

[64] G. Frey. Links between stable elliptic curves and certain diophantine equations. *Annales Universitatis Saraviensis*, **1**, 1–40, 1986.

[65] I. Gaál. Power integral bases in orders of families of quartic fields. *Publ. Math. Debrecen*, **42**, 253–263, 1993.

[66] I. Gaál. Computing elements of given index in totally complex cyclic sextic fields. *J. Symbolic Computation*, **20**, 61–69, 1995.

[67] I. Gaál. Computing all power integral bases in orders of totally real cyclic sextic number fields. *Math. Comp.*, **65**, 801–822, 1996.

[68] I. Gaál. Power integral bases in composites of number fields. To appear: *Canad. Math. Bull.*

[69] I. Gaál, A. Pethő, and M. Pohst. On the resolution of index form equations in biquadratic number fields II. *J. Number Theory*, **38**, 35–51, 1991.

[70] I. Gaál, A. Pethő, and M. Pohst. On the resolution of index form equations in quartic number fields. *J. Symbolic Computation*, **16**, 563–584, 1993.

[71] I. Gaál, A. Pethő, and M. Pohst. Simultaneous representation of integers by a pair of ternary quadratic forms – with an application to index form equations in quartic number fields. *J. Number Theory*, **57**, 90–104, 1996.

[72] I. Gaál and M. Pohst. On the resolution of index form equations in sextic fields with an imaginary quadratic subfield. *J. Symbolic Computation*, **22**, 425–434, 1996.

[73] I. Gaál and M. Pohst. Power integral bases in a parametric family of totally real cyclic quintics. To appear: *Math. Comp.*

[74] I. Gaál and N. Schulte. Computing all power integral bases of cubic fields. *Math. Comp.*, **53**, 689–696, 1989.

[75] J. Gebel, E. Herrmann, A. Pethő, and H.G. Zimmer. Computing all $S$–integral points on elliptic curves. Preprint.

[76] J. Gebel, A. Pethő, and H.G. Zimmer. Computing $S$–integral points on elliptic curves. In [**33**], pages 157–171.

[77] J. Gebel, A. Pethő, and H.G. Zimmer. Computing integral points on elliptic curves. *Acta. Arith.*, **68**, 171–192, 1994.

[78] J. Gebel, A. Pethő, and H.G. Zimmer. On Mordell's equation. To appear: *Comp. Math.*

[79] J. Gebel and H.G. Zimmer. Computing the Mordell–Weil group of an elliptic curve over $\mathbb{Q}$. In H. Kisilevsky and M. Ram. Murty, editors, *Elliptic Curves and Related Topics*. CRM Proceedings and Lecture Notes Volume 4, AMS ., 1994.

[80] A.M.W. Glass, D.B. Meronk, T. Okada, and R.P. Steiner. A small contribution to Catalan's equation. *J. Number Theory*, **47**, 131–137, 1994.

[81] D.M. Gordon and D.R. Grant. Computing the Mordell–Weil rank of Jacobians of curves of genus two. *Trans. AMS*, **337**, 807–824, 1993.

[82] D.R. Grant. Formal groups in genus two. *J. Reine Angew. Math.*, **411**, 96–121, 1990.

[83] M.J. Greenberg. *Lectures on Forms in Many Variables.* W.A. Benjamin, 1969.

[84] R. Gross and J.H. Silverman. $S$–integer points on elliptic curves. *Pacific J. Maths*, **167**, 263–288, 1995.

[85] R.K. Guy. The Ochoa curve. *Crux. Mathematicorum*, **16**, 65–69, 1990.

[86] K. Győry. Sur les polynomes á coefficients entiers et de discriminant donné, III. *Publ. Math. Debrecen*, **23**, 141–165, 1976.

[87] K. Győry. On polynomials with integer coefficients and given discriminant IV. *Publ. Math. Debrecen*, **25**, 155–167, 1978.

[88] K. Győry. On polynomials with integer coefficients and given discriminant V , $p$–adic generalizations. *Acta. Math.*, **32**, 175–190, 1978.

[89] K. Győry. On the greatest prime factors of decomposable forms at integer points. *Ann. Acad. Sci. Fenn Ser. A.I Math.*, **4**, 341–355, 1978/9.

[90] K. Győry. On the number of solutions of linear equations in units of an algebraic number field. *Comm. Math. Helvetici*, **54**, 585–600, 1979.

[91] K. Győry. Explicit upper bounds for the solutions of some diophantine equations. *Ann. Acad. Sci. Fenn. Ser. A.I Math.*, **5**, 3–12, 1980.

[92] K. Győry. On certain graphs composed of algebraic integers of a number field and their applications I. *Publ. Math. Debrecen*, **27**, 229–242, 1980.

[93] K. Győry. On the representation of integers by decomposable forms in several variables. *Publ. Math. Debrecen*, **28**, 89–98, 1981.

[94] K. Győry and Z. Papp. Effective estimates for the integer solutions of norm form and discriminant form equations. *Publ. Math. Debrecen*, **25**, 311–325, 1978.

[95] L. Hajdu and T. Herendi. Explicit bounds for the solutions of elliptic equations with rational coefficients. To appear: *J. Symbolic Computation.*

[96] G.H. Hardy and E.M. Wright. *An Introduction to the Theory of Numbers.* Oxford University Press, Oxford, 1954.

[97] R. Hartshorne. *Algebraic Geometry.* Springer–Verlag, GTM 52, 1977.

[98] C. Hermite. Sur l'introduction des variables continues dans la theorie des nombres. *J. Reine Angew. Math.*, **41**, 191–216, 1851.

[99] C. Heuberger. On a family of quintic Thue equations. To appear: *J. Symbolic Computation.*

[100] D. Hilbert. *Theory of Algebraic Invariants*. Cambridge University Press, 1993.

[101] D. Husemoller. *Elliptic Curves*. Springer–Verlag, GTM 111, 1986.

[102] K. Inkeri. On Catalan's problem. *Acta. Arith*, **9**, 285–290, 1964.

[103] K. Inkeri. On Catalan's conjecture. *J. Number Th.*, **34**, 142–152, 1990.

[104] G. Julia. Étude sur les formes binaires non quadratiques. *Mem. Acad. Sci. l'Inst. France*, **55**, 1–293, 1917.

[105] A. Knapp. *Elliptic Curves*. Princeton Univ. Press, 1992.

[106] Chao Ko. On the diophantine equation $x^2 = y^n + 1$, $xy \neq 0$. *Sci. Sinica*, **14**, 457–460, 1965.

[107] N. Koblitz. *Introduction to elliptic curves and modular forms*. Springer–Verlag, GTM 97, 1984.

[108] N. Koblitz. *P–adic Numbers, P–adic Analysis and Zeta Functions*. Springer–Verlag, GTM 58, 1984.

[109] J.L. Lagrange. Reserches d'arithmetique. *Nouv. Mem. Acad. Berlin*, pages 265–312, 1773.

[110] S. Lang. Diophantine approximation on toruses. *Amer. J. Math.*, **86**, 521–533, 1964.

[111] S. Lang. *Elliptic Curves: Diophantine Analysis*. Springer–Verlag, 1978.

[112] S. Lang. *Fundamentals of Diophantine Geometry*. Springer–Verlag, 1983.

[113] S. Lang. *Algebraic Number Theory*. Springer–Verlag, GTM 110, 1986.

[114] V.A. Lebesgue. Sur l'impossibilité en nombres entiers de l'équation $x^m = y^2 + 1$. *Nouv. Ann. Math*, **9**, 178–181, 1850.

[115] A.K. Lenstra. Factoring polynomials over algebraic number fields. In *Proc EUROCAL83*, LNCS 162, pages 245–254. Springer–Verlag, 1983.

[116] A.K. Lenstra and H.W. Lenstra, editors. *The Development of the Number Field Sieve*. Springer–Verlag, LNM 1554, 1993.

[117] A.K. Lenstra, H.W. Lenstra, and L. Lovász. Factoring polynomials with rational coefficients. *Math. Ann.*, **261**, 515–534, 1982.

[118] H.W. Lenstra. Euclidean number fields of large degree. *Invent. Math.*, **38**, 237–254, 1977.

[119] G. Lettl and A. Pethő. Complete solution of a family of quartic Thue equations. *Hamburger Abhandlungen*, **65**, 365–383, 1995.

[120] G. Lettl, A. Pethő, and P. Voutier. Simple families of Thue equations. To appear: *Trans. AMS.*

[121] A. Leutbecher and G. Niklasch. On cliques of exceptional units and Lenstra's construction of Euclidean fields. In H.P. Schlickewei and E. Wirsing, editors, *Number Theory, Proc Jour. Arith., Ulm 1987*, pages 150–178. Springer–Verlag, LNM 1380, 1989.

[122] LiDIA Group. LiDIA v1.3 – a library for computational number theory. *TH Darmstadt*, 1997.

[123] W. Ljunggren. On the diophantine equation $y^2 - k = x^3$. *Acta. Arith.*, **8**, 451–463, 1963.

[124] K. Mahler. Zur Approximation algebraischer Zahlen, 1: Über den grőssten Primteiler binărer Formen. *Math. Ann.*, **107**, 691–730, 1933.

[125] J. Manin. The $p$–torsion of elliptic curves is uniformly bounded. *Isv. Akad. Nauk. SSSR Ser. Mat., Amer. Math. Soc. Transl.*, **33**, 433–438, 1969.

[126] J.R. Merriman. *Binary forms and the reduction of curves*. Ph.D. thesis, Oxford University, 1970.

[127] J.R. Merriman, S. Siksek, and N.P. Smart. Explicit 4–descents on an elliptic curve. *Acta. Arith.*, **77**, 385–404, 1996.

[128] J.R. Merriman and N.P. Smart. The calculation of all algebraic integers of degree 3 with discriminant a product of powers of 2 and 3 only. *Publ. Math. Debrecen*, **43**, 195–205, 1993.

[129] J.R. Merriman and N.P. Smart. Curves of genus 2 with good reduction away from 2 with a rational Weierstrass point. *Proc. Camb. Phil. Soc.*, **114**, 203–214, 1993.

[130] M. Mignotte. Verification of a conjecture of E. Thomas. *J. Number Theory*, **44**, 172–177, 1993.

[131] M. Mignotte. Sur l'équation de Catalan, II. *Theoretical Computer Science*, **123**, 145–149, 1994.

[132] M. Mignotte, A. Pethő, and R. Roth. Complete solutions of a family of quartic Thue and index form equations. *Math. Comp.*, **65**, 341–354, 1996.

[133] M. Mignotte and B.M.M. de Weger. On the diophantine equations $x^2 + 74 = y^5$ and $x^2 + 86 = y^5$. *Glasgow Math. Journal*, **38**, 77–85, 1996.

[134] M. Mignotte and Y.Roy. Catalan's equation has no new solution with either exponent less than 10651. *Experimental Math.*, **4**, 259–268, 1995.

[135] M. Mignotte and Y.Roy. Minorations pour l'équation de Catalan. *C.R. Acad. Sci. Paris*, **324**, 377–380, 1997.

[136] J.S. Milne. Jacobian varieties. In G. Cornell and J.H. Silverman, editors, *Arithmetic Algebraic Geometry*, pages 167–212. Springer–Verlag, 1986.

[137] R.A. Mollin, editor. *Number Theory: Proc. First Conference of the Canadian Number Theory Association, 1988.* W. de Gruyter, Berlin, 1990.

[138] L.J. Mordell. *Diophantine Equations.* Academic Press, 1969.

[139] T. Nagell. Des équations indéterminées $x^2 + x + 1 = y^n$ et $x^2 + x + 1 = 3y^n$. *Nordsk. Mat. Forenings Skr., ser. I, nr. 2*, 1919.

[140] T. Nagell. Sur une propriété des unités d'un corps algébrique. *Arkiv f. Matem.*, **5**, 343–356, 1964.

[141] G. Niklasch. Family portraits of exceptional units. Preprint.

[142] G. Niklasch and N.P. Smart. Exceptional units in a family of quartic number fields. *Math. Comp.*, **67**, 759–772, 1998.

[143] A.M. Odlyzko. The rise and fall of knapsack cryptosystems. In C. Pomerance, editor, *Cryptology and computational number theory*, pages 75–88. Proc. Symp. Applied Maths Vol. 42, 1990.

[144] A. Pethő. Computational methods for the resolution of diophantine equations. In [**137**], pages 479–492.

[145] A. Pethő. Full cubes in the Fibonacci sequence. *Publ. Math. Debrecen*, **30**, 117–127, 1983.

[146] A. Pethő. On the resolution of Thue inequalities. *J. Symbolic Computation*, **4**, 103–109, 1987.

[147] A. Pethő. Complete solutions to families of quartic Thue equations. *Math. Comp.*, **57**, 777–798, 1991.

[148] A. Pethő. Complexity investigations on decomposable form equations. *Publ. Math. Debrecen*, **39**, 163–169, 1991.

[149] A. Pethő and R. Schulenberg. Effektives Lösen von Thue Gleichungen. *Publ. Math. Debrecen*, **34**, 189–196, 1987.

[150] A. Pethő and R.F. Tichy. On two–parametric quartic families of diophantine problems. To appear: *J. Symbolic Computation.*

[151] A. Pethő and B.M.M. de Weger. Products of prime powers in binary recurrence sequences I. The hyperbolic case, with an application to the generalized Ramanujan–Nagell equation. *Math. Comp.*, **47**, 713–727, 1986.

[152] A. Pethő, M. Pohst, H.C. Williams, and H.G. Zimmer, editors. *Computational Number Theory.* Walter de Gruyter, 1991.

[153] M. Pohst. On the computation of number fields of small discriminants including the minimum discriminants of sixth degree fields. *J. Number Theory*, **14**, 99–117, 1982.

[154] M. Pohst. *Computational Algebraic Number Theory.* Birkäuser, DMV Seminar Band 21, 1993.

[155] M. Pohst and H. Zassenhaus. *Algorithmic Algebraic Number Theory.* Cambridge University Press, 1989.

[156] B. Poonen. Computational aspects of curves of genus at least 2. In [**33**], pages 283–306.

[157] G. Rémond and F. Urfels. Approximations diophantiennes de logarithmes elliptiques $p$-adiques. *J. Number Th.*, **57**, 133–169, 1996.

[158] K. Ribet. On modular representations of $Gal(\overline{\mathbb{Q}}/\mathbb{Q})$ arising from modular forms. *Invent. Math.*, **100**, 431–476, 1990.

[159] J.H. Rickert. Simultaneous rational approximation and related diophantine equations. *Proc. Camb. Phil. Soc.*, **113**, 461–472, 1993.

[160] H.E. Rose. *A Course in Number Theory.* Oxford Science Publications, 1988.

[161] G. Salmon. *Modern Higher Algebra.* Hodges, Foster and Co., 1876.

[162] E.F. Schaefer. 2–descent on the Jacobians of hyperelliptic curves. *J. Number Theory*, **51**, 219–232, 1995.

[163] E.F. Schaefer. Class groups and Selmer groups. *J. Number Theory*, **56**, 79–114, 1996.

[164] C.P. Schnorr. A more efficient algorithm for lattice basis reduction. *J. Algorithms*, **9**, 47–62, 1988.

[165] W. Schwartz. A note on Catalan's equation. *Acta. Arith.*, **72**, 277–279, 1995.

[166] P. Serf. *The rank of elliptic curves over real quadratic number fields of class number 1.* Ph.D. thesis, Universität des Saarlandes, 1995.

[167] T. Shorey and R. Tijdeman. *Exponential Diophantine Equations.* Cambridge University Press, 1986.

[168] C.L. Siegel. Über einige Anwendungen diophantischer Approximationen. *Abh. Preuss. Akad. Wiss.*, pages 1–41, 1929.

[169] S. Siksek. Infinite descent on elliptic curves. *Rocky Mountain Journal of Maths*, **25**, 1501–1538, 1995.

[170] S. Siksek. Sieving for rational points on hyperelliptic curves. Preprint

[171] S. Siksek and N.P. Smart. On the complexity of computing the 2–Selmer group of an elliptic curve. *Glasgow Math. Journal.*, **39**, 251–258, 1997.

[172] J.H. Silverman. *The Arithmetic of Elliptic Curves.* Springer–Verlag, GTM 106, 1986.

[173] J.H. Silverman. Computing heights on elliptic curves. *Math. Comp.*, **51**, 339–358, 1988.

[174] J.H. Silverman. The difference between the Weil height and the canonical height on elliptic curves. *Math. Comp.*, **55**, 723–743, 1990.

[175] J.H. Silverman. Computing canonical heights with little (or no) factorization. To appear: *Math. Comp.*

[176] J.H. Silverman and J.T. Tate. *Rational Points on Elliptic Curves.* Springer–Verlag, 1992.

[177] SIMATH Group. SIMATH Group – a computer algebra system for algorithmic number theory. *Univ. Saarbrueken*, 1997.

[178] N.P. Smart. Solving a quartic discriminant form equation. *Publ. Math. Debrecen*, **43**, 29–39, 1993.

[179] N.P. Smart. $S$–integral points on elliptic curves. *Proc. Camb. Phil. Soc.*, **116**, 391–399, 1994.

[180] N.P. Smart. The solution of triangularly connected decomposable form equations. *Math. Comp.*, **64**, 819–840, 1995.

[181] N.P. Smart. Solving discriminant form equations via unit equations. *J. Symbolic Computation*, **21**, 367–374, 1996.

[182] N.P. Smart. $S$–unit equations, binary forms and curves of genus 2. *Proc. London Math. Soc.*, **75**, 271–307, 1997.

[183] N.P. Smart. Thue and Thue–Mahler equations over rings of integers. To appear: *J. LMS*.

[184] N.P. Smart. Determining the small solutions to $S$–unit equations. Preprint.

[185] N.P. Smart and N.M. Stephens. Integral points on elliptic curves over number fields. *Proc. Camb. Phil. Soc.*, **122**, 9–16, 1997.

[186] R.P. Steiner. On Mordells equation $y^2 - k = x^3$: A problem of Stolarsky. *Math. Comp.*, **46**, 703–714, 1986.

[187] I.N. Stewart and D.O. Tall. *Algebraic Number Theory*. Chapman and Hall, 1979.

[188] R.J. Stroeker. On Thue equations associated with certain quartic number fields. In [**152**], pages 313–319.

[189] R.J. Stroeker. On the sum of consecutive cubes being a perfect square. *Comp. Math.*, **97**, 295–307, 1995.

[190] R.J. Stroeker and N. Tzanakis. On the application of Skolem's $p$–adic method to the solution of Thue equations. *J. Number Theory*, **29**, 166–195, 1988.

[191] R.J. Stroeker and N. Tzanakis. Solving elliptic diophantine equations by estimating linear forms in elliptic logarithms. *Acta. Arith.*, **67**, 177–196, 1994.

[192] R.J. Stroeker and B.M.M. de Weger. On elliptic diophantine equations that defy Thue–the case of the Ochoa curve, *Exp. Math.*, **3**, 209–220, 1994.

[193] R.J. Stroeker and B.M.M. de Weger. On a quartic diophantine equation. *Proc. Edin. Math. Soc.*, **39**, 97–114, 1996.

[194] J.T. Tate. The arithmetic of elliptic curves. *Invent. Math*, **23**, 179–206, 1974.

[195] E. Thomas. Complete solutions to a family of cubic diophantine equations. *J. Number Theory*, **34**, 235–250, 1990.

[196] A. Thue. Über Annäherungswerte algebraischer Zahlen. *J. Reine Angew Math*, **135**, 284–305, 1909.

[197] R. Tijdeman. On the equation of Catalan. *Acta. Arith.*, **29**, 197–209, 1976.

[198] J. Top. *Hecke L–series related with algebraic cycles or with Siegel modular forms*. Ph.D. thesis, Utrecht, 1989.

[199] N. Tzanakis. The diophantine equation $x^3 + 3y^3 = 2^n$. *J. Number Theory*, **15**, 376–387, 1982.

[200] N. Tzanakis. The diophantine equation $x^3 - 3xy^2 - y^3 = 1$ and related equations. *J. Number Theory*, **18**, 192–205, 1984.

[201] N. Tzanakis. Solving elliptic diophantine equations by estimating linear forms in elliptic logarithms. The case of quartic equations. *Acta. Arith.*, **75**, 165–190, 1996.

[202] N. Tzanakis and B.M.M. de Weger. On the practical solution of the Thue equation. *J. Number Theory*, **31**, 99–132, 1989.

[203] N. Tzanakis and B.M.M. de Weger. Solving a specific Thue–Mahler equation. *Math. Comp.*, **57**, 799–815, 1991.

[204] N. Tzanakis and B.M.M. de Weger. How to explicitly solve a Thue–Mahler equation. *Comp. Math.*, **84**, 223–288, 1992.

[205] P.M. Voutier. On the number of $S$–integral solutions to $y^m = f(x)$. Preprint.

[206] I. Wakabayashi. On a family of quartic Thue inequalities. *J. Number Theory*, **66**, 70–84, 1997.

[207] B.M.M. de Weger. Solving exponential diophantine equations using lattice basis reduction algorithms. *J. Number Theory*, **26**, 325–367, 1987.

[208] B.M.M. de Weger. *Algorithms For Diophantine Equations*. Centre For Mathematics And Computer Science Amsterdam, 1989. CWI–Tract.

[209] B.M.M. de Weger. A hyperelliptic diophantine equation related to imaginary quadratic number fields with class number 2. *J. Reine Angew. Math.*, **427**, 137–156, 1992.

[210] B.M.M. de Weger. A curious property of the eleventh Fibonacci number. *Rocky Mountain Journal Of Maths*, **25**, 977–994, 1995.

[211] B.M.M. de Weger. A Thue equation with quadratic integers as variables. *Math. Comp.*, **64**, 855–861, 1995.

[212] B.M.M. de Weger. A binomial diophantine equation. *Quarterly J. Math. Oxford*, **47**, 221–231, 1996.

[213] B.M.M. de Weger. One diophantine equation. Preprint.

[214] B.M.M. de Weger. Padua and Pisa are exponentially far apart. To appear: *Publ. Math. Debrecen.*

[215] E.T. Whittaker and G.N. Watson. *A course in modern analysis.* Camb. Univ. Press, 1927.

[216] K. Wildanger. *Über das Lösen von Einheiten- und Indexformgleichungen in algebraischen Zahlkörpern mit einer Anwendung auf die Bestimmung aller ganzen Punkte einer Mordellschen Kurve.* Ph.D. thesis, Technischen Universität Berlin, 1997.

[217] A. Wiles. Modular elliptic curves and Fermat's Last Theorem. *Annals of Maths*, **142**, 443–551, 1995.

[218] K.R. Yu. Linear forms in $p$–adic logarithms. *Acta. Arith.*, **53**, 107–186, 1989.

[219] D. Zagier. Large integral points on elliptic curves. *Math. Comp.*, **48**, 425–436, 1987.

[220] M.E. Zieve. *Cycles of polynomial mappings.* Ph.D. thesis, University of California, Berkeley, 1996.

[221] H.G. Zimmer. A limit formula for the canonical height of an elliptic curve and its applications to height computations. In [**137**], pages 641–659.

[222] H.G. Zimmer. On the difference between the Weil height and the Neron–Tate height. *Math. Z.*, **147**, 35–51, 1976.

# Index

For EU product safety concerns, contact us at Calle de José Abascal, 56–1°, 28003 Madrid, Spain or eugpsr@cambridge.org.

www.ingramcontent.com/pod-product-compliance
Ingram Content Group UK Ltd.
Pitfield, Milton Keynes, MK11 3LW, UK
UKHW042209080726
066UK00007B/323